Management and Industrial Engineering

Series Editor

J. Paulo Davim, Department of Mechanical Engineering, University of Aveiro, Aveiro, Portugal

D1806696

This series fosters information exchange and discussion on management and industrial engineering and related aspects, namely global management, organizational development and change, strategic management, lean production, performance management, production management, quality engineering, maintenance management, productivity improvement, materials management, human resource management, workforce behavior, innovation and change, technological and organizational flexibility, self-directed work teams, knowledge management, organizational learning, learning organizations, entrepreneurship, sustainable management, etc. The series provides discussion and the exchange of information on principles, strategies, models, techniques, methodologies and applications of management and industrial engineering in the field of the different types of organizational activities. It aims to communicate the latest developments and thinking in what concerns the latest research activity relating to new organizational challenges and changes world-wide. Contributions to this book series are welcome on all subjects related with management and industrial engineering. To submit a proposal or request further information, please contact Professor J. Paulo Davim, Book Series Editor, pdavim@ua.pt

More information about this series at https://link.springer.com/bookseries/11690

Carolina Machado · J. Paulo Davim
Editors

Sustainability and Intelligent Management

 Springer

Editors
Carolina Machado ⓘ
Department of Management, School
of Economics and Management
University of Minho
Braga, Portugal

J. Paulo Davim ⓘ
Department of Mechanical Engineering
University of Aveiro
Aveiro, Portugal

ISSN 2365-0532   ISSN 2365-0540 (electronic)
Management and Industrial Engineering
ISBN 978-3-030-98038-2   ISBN 978-3-030-98036-8 (eBook)
https://doi.org/10.1007/978-3-030-98036-8

This Springer imprint is published by the registered company Springer Nature Switzerland AG
The registered company address is: Gewerbestrasse 11, 6330 Cham, Switzerland

Preface

Sustainability and intelligent management are two critical concepts every day present in the organizations present lives. According to the Environmental Science.Org (2020), sustainability is understood as a broad discipline that gives to everybody insights into most aspects of the human world from business to technology, to environment, and the social sciences. Its main focus consists in answering to the current needs without compromising the reach of needs for future generations. This means that instead of looking at the short term, the immediate profits resulting from their activity, organizations should look to exploit the interactions between the economic, social, environmental, and human components in a long-term perspective.

Intelligent management, by its side, looks to transfer to organizations the knowledge, know-how and current practices, necessary to overcome the structured management systems characteristic of the most classic organizations, shifting toward a network and system-based perspective to more effectively answer to the needs and challenges of today's organizations. Indeed, organizations closed in on themselves cannot survive in today's competitive markets. In a context where digital transformation puts constant pressure on the way organizations operate, intelligent management is seen as a systemic thinking approach that nowadays managers and leaders can use to define the organization strategies and procedures from the initial idea to its implementation.

Joined, sustainability and intelligent management are of huge relevance in the management of competitive and social' responsible organizations. In other words, intelligent management can be seen as a critical tool to help organizations in the implementation and control of sustainable practices. At this level, organizations need to combine different types of knowledge, namely in what concerns digitalization, human behavior, new information technologies, as well as business and organizational understanding, in order to be more competitive and more than that, responsible in response to the growing demands of the environment in which they operate.

Sustainability and intelligent management are thus understood as two key concepts that when used together, will contribute and help organizations to better answer to the challenges and changes characteristic of current markets and the environment in which they are involved, through the creation and implementation of sustainable and intelligent business models. Applying intelligent management to the design/identification, analysis, management, development, and reporting of sustainable information and concerns, will allow organizations to deeply assess all their information (and not just, or even, excluding financial information), contributing, by this way, to achieve their main objectives.

Conscious of the importance of these two concepts in the organizational context, with the present book we can find different contributions that allow us to better understand the main contributions of intelligent management theories and models in a more effective management of sustainable business practices in competitive, profitable and social' responsible organizations.

Given its characteristics, intelligent management is a critical tool that nowadays managers and engineers need to know and develop in order to better find, obtain and make decisions which best cater for their and the organization interests on the path to the necessary sustainability.

Attentive to its importance in today's organizations, the present book entitled *Sustainability and Intelligent Management* looks to communicate the latest developments and thinking on the sustainability and intelligent management subjects worldwide.

Seeking cultural and geographic diversity in studies related to *Sustainability and Intelligent Management* issues, this book, organized in seven chapters, begins to present "High-Quality Employment Relationships: Sustainable Management Through a Psychological Contract Perspective", covers "Designing an Interactionist Approach for Sustainable HRM: Might the Borders and Overlaps with Internal Communication Lead to a Common End?", discusses "The Recruitment and Selection Processes Influence on the Access of Immigrants to the Labor Market", focuses on "Can Facebook Data Predict the Level of Sustainable Development in EU-27?", deals with "The Influence of Social Participation for Sustainable Against Regression: A Critical Review of Brazilian Environmental Public Policies in Light of the Environmental Justice", speaks about "Analysis of Food Loss and Waste for the European Countries in the Context of Sustainable Development", and finally contains information about "Sustainability, Innovation, and Diversification in the Spanish Frozen Food Industry: A Financial Analysis".

This book is designed to increase the knowledge and effectiveness of all those involved in the areas of management and sustainability whether in the profit or non-profit sectors, or in the public or private sectors. Indeed, given the high potentialities of these issues to managers and engineers, academics, researchers, and other professionals, the interest in this subject, at the present and more and more in the future, is evident for many types of organizations, namely, important Institutes and Universities in the world.

The Editors acknowledge their gratitude to Springer for this opportunity and for their professional support. Finally, we would like to thank all chapter authors for their interest and availability to work on this project.

Braga, Portugal   Carolina Machado
                                                                carolina@eeg.uminho.pt

Aveiro, Portugal   J. Paulo Davim
                                                                       pdavim@ua.pt

Contents

High-Quality Employment Relationships: Sustainable
Management Through a Psychological Contract Perspective   1
Johannes M. Kraak and Yannick Griep

Designing an Interactionist Approach for Sustainable HRM:
Might the Borders and Overlaps with Internal Communication
Lead to a Common End? 21
Daniel Roque Gomes and Maria João Santos

Recruitment and Selection Processes Influence on the Access
of Immigrants to the Labor Market 41
Raquel Cerqueira Gonçalves and Carolina Feliciana Machado

Can Facebook Data Predict the Level of Sustainable Development
in EU-27? 61
Marius Constantin, Jean-Vasile Andrei, Drago Cvijanovic,
and Teodor Sedlarski

The Influence of Social Participation for Sustainable Against
Regression: A Critical Review of Brazilian Environmental Public
Policies in Light of the Environmental Justice 107
Paulo Santos de Almeida and Vitor Calandrini de Araujo

Analysis of Food Loss and Waste for the European Countries
in the Context of Sustainable Development 119
Adrian Stancu

Sustainability, Innovation and Diversification in the Spanish
Frozen Food Industry: A Financial Analysis 149
Félix Puime-Guillén, Raisa Pérez-Vas, and Raquel Fernández-González

Index 161

Editors and Contributors

About the Editors

Carolina Machado received her Ph.D. degree in Management Sciences (Organizational and Policies Management area/Human Resources Management) from the University of Minho in 1999, Master degree in Management (Strategic Human Resource Management) from the Technical University of Lisbon in 1994, and Degree in Business Administration from the University of Minho in 1989. Teaching in the Human Resources Management subjects since 1989 at University of Minho, she is since 2004 Associated Professor, with experience and research interest areas in the field of Human Resource Management, International Human Resource Management, Human Resource Management in SMEs, Training and Development, Emotional Intelligence, Management Change, Knowledge Management and Management/HRM in the Digital Age/Business Analytics. She is Head of the Human Resources Management Work Group at the School of Economics and Management at the University of Minho, Coordinator of Advanced Training Courses at the Interdisciplinary Centre of Social Sciences, Member of the Interdisciplinary Centre of Social Sciences (CICS.NOVA.UMinho), University of Minho, as well as Chief Editor of the *International Journal of Applied Management Sciences and Engineering* (IJAMSE), Guest Editor of journals, books editor and books series editor, as well as reviewer in different international prestigious journals. In addition, she has also published both as editor/co-editor and as author/co-author of several books, book chapters, and articles in journals and conferences.

J. Paulo Davim is a Full Professor at the University of Aveiro, Portugal. He is also distinguished as honorary professor in several universities/colleges/institutes in China, India, and Spain. He received his Ph.D. degree in Mechanical Engineering in 1997, M.Sc. degree in Mechanical Engineering (materials and manufacturing processes) in 1991, Mechanical Engineering degree (5 years) in 1986, from the University of Porto (FEUP), the Aggregate title (Full Habilitation) from the University of Coimbra in 2005 and the D.Sc. (Higher Doctorate) from London Metropolitan

University in 2013. He is a Senior Chartered Engineer by the Portuguese Institution of Engineers with an M.B.A. and Specialist titles in Engineering and Industrial Management as well as in Metrology. He is also Eur Ing by FEANI-Brussels and Fellow (FIET) of IET-London. He has more than 35 years of teaching and research experience in Manufacturing, Materials, Mechanical, and Industrial Engineering, with special emphasis in Machining & Tribology. He has also interest in Management, Engineering Education, and Higher Education for Sustainability. He has guided large numbers of postdoc, Ph.D., and master's students as well as has coordinated and participated in several financed research projects. He has received several scientific awards and honors. He has worked as an evaluator of projects for ERC-European Research Council and other international research agencies as well as examiner of Ph.D. thesis for many universities in different countries. He is the Editor in Chief of several international journals, Guest Editor of journals, books Editor, book Series Editor and Scientific Advisory for many international journals and conferences. Presently, he is an Editorial Board member of 30 international journals and acts as a reviewer for more than 100 prestigious Web of Science journals. In addition, he has also published as editor (and co-editor) more than 200 books and as author (and co-author) more than 15 books, 100 book chapters, and 500 articles in journals and conferences (more than 300 articles in journals indexed in Web of Science core collection/h-index 61+ /12500+ citations, SCOPUS/h-index 66+ /15500+ citations, Google Scholar/h-index 85+/25500+ citations). He has been listed in World's Top 2% Scientists by Stanford University study.

Contributors

Jean-Vasile Andrei Petroleum-Gas University of Ploiesti, Ploiesti, Romania; National Institute for Economic Research 'Costin C. Kiritescu', Romanian Academy, Bucharest, Romania

Marius Constantin Bucharest University of Economic Studies, Bucharest, Romania

Drago Cvijanovic University of Kragujevac, Kragujevac, Serbia

Paulo Santos de Almeida Programa de Pós-graduação em Sustentabilidade, Escola de Artes, Ciências e Humanidades, Universidade de São Paulo, Rua Arlindo Bettio, 1000, São Paulo, Brasil

Vitor Calandrini de Araujo Programa de Pós-graduação em Sustentabilidade, Escola de Artes, Ciências e Humanidades, Universidade de São Paulo, São Paulo, Brasil

Raquel Fernández-González Department of Applied Economics, ERENEA-ECOBAS, University of Vigo, Vigo, Spain

Daniel Roque Gomes Polytechnic Institution of Coimbra, School of Education, R. Misericórdia Lagar dos Cortiços—S. Martinho do Bispo, 3045-093 Coimbra, Portugal;
ICNOVA, Instituto de Comunicação da NOVA, FSCH da Universidade Nova de Lisboa, Avenida de Berna, 1069-061 Lisboa, Portugal

Raquel Cerqueira Gonçalves Department of Management, School of Economics and Management, University of Minho, Braga, Portugal

Yannick Griep Behavioural Science Institute, Radboud University, 6525 Nijmegen, The Netherlands;
Stress Research Institute, Stockholm University, 114 19 Stockholm, Sweden

Johannes M. Kraak KEDGE Business School, Bordeaux, France

Carolina Feliciana Machado Department of Management, School of Economics and Management, University of Minho, Braga, Portugal;
Interdisciplinary Centre of Social Sciences (CICS.NOVA.UMinho), University of Minho, Braga, Portugal

Félix Puime-Guillén Department of Business, University of A Coruña, A Coruña, Spain

Raisa Pérez-Vas Department of Financial Economics and Accounting, IC2-ECOBAS, University of Vigo, Vigo, Spain

Maria João Santos Lisbon School of Economics and Management, SOCIUS—Research Center in Economic and Organizational Sociology, University of Lisbon, Lisbon, Portugal

Teodor Sedlarski St Kliment Ohridski University of Sofia, Sofia, Bulgaria

Adrian Stancu Petroleum-Gas University of Ploiesti, Ploiesti, Romania

High-Quality Employment Relationships: Sustainable Management Through a Psychological Contract Perspective

Johannes M. Kraak and Yannick Griep

Abstract The literature on sustainable management has seen a stark increase in publications in the last few years. A considerable number of papers were written on models or specific human resource practices that aim to provide clarity on what sustainable human resources entail. However, many publications seem to show that sustainable management focuses primarily on employer-driven outcomes and external stakeholders. In this chapter we therefore focus on employees and high-quality employment relationships through the lens of the psychological contract. We propose that by focusing on developing a high-quality psychological contract, companies will select the right contract terms in the form of sustainable human resource practices so as to realize the desired outcomes such as employee commitment and retention. After a short introduction on sustainability, we discuss psychological contracts and zoom in on three specific areas in which companies can make a real difference in generating high-quality relationships, thereby putting in place an employee-centered sustainable human resource management.

Keywords Sustainable management · Sustainable HR · Social exchange · Psychological contract

1 Introduction

In this chapter we will explore sustainable management through the lens of the employment relationship, which is an employee's primary source of work-related rights and benefits (International Labour Organization, 2021). More specifically, we will focus on psychological contract research, which aims to better understand the employment relationship (Griep et al., 2019). We believe that a high-quality

---

J. M. Kraak (✉)
KEDGE Business School, 33405 Bordeaux, France
e-mail: johannes.kraak@kedgebs.com

Y. Griep
Behavioural Science Institute, Radboud University, 6525 Nijmegen, The Netherlands

Stress Research Institute, Stockholm University, 114 19 Stockholm, Sweden psychological contract is synonym to sustainable management as a well-functioning employment relationship, in which employees feel respected, is likely to trigger positive outcomes such as a strong desire to stay with the company. Our goal here is not to provide an exhaustive overview of all the possible aspects of sustainable (human resource) management because other chapters in this book will certainly do so already. Instead, we focus on the relationship between employees and their employers and identify areas where the psychological contract plays an integral part in creating a high-quality and thus sustainable employment relationship. In this first section we will start with a short discussion on the hype around sustainability and how this relates to the everyday reality for employees. We then discuss the sustainable management literature and the place of employees in this field before providing the reader with an introduction on high-quality employment relationships.

1.1   The Sustainability Hype

References to sustainability have become omnipresent in today's economy. Sustainability is used to describe organizations, their products, supply chain partners, employees, and even customers. We are led to believe that sustainability is all around. A basic count of the number of times that a generic sustainability-like reference is made on a random companies' website says it all. The term sustainability seems to have replaced the use of *greenwashing* (i.e., engaging in "symbolic communications of environmental issues without substantially addressing them in action," Walker & Wan, 2012, p. 227), which has been heavily criticized over the last few years due to the fact that it seems opportunistic and primarily aimed at masking negative environmental aspects of business. Communicating more on sustainability in a more holistic corporate social responsibility (CSR) setting seems like a good alternative to an increased fixation on mostly environmental aspects in faraway places because it allows companies to talk about broader values and beliefs at home. Communicating CSR-values is important in building a stronger brand image and leads to increased attractiveness of the organization among job seekers as well as employee retention (e.g., Dögl & Holtbrügge, 2014; Puncheva-Michelotti et al., 2018). As a result, many companies have embraced the sustainability trend and are more actively pushing CSR-oriented communication when attracting new employees. For instance, a visit to the employment section of the Ikea France website in August of 2021 started off with a description of Ikea's diverse workforce, working in an open and inclusive environment that aims to continuously improve and increase sustainability for all. The CSR-orientation in corporate communication is taking on such proportions that the term *CSR-washing* (i.e., organizations trying to come across as more socially responsible than they actually are) recently emerged in the literature (Ginder et al., 2021). The question arises if these CSR-related statements are representative for the quality of the employment relationship that current and future employees of these companies will experience. As with greenwashing it seems that many organizations who actively use sustainability references have a dark side too, potentially undermining the credibility of what these companies set out to achieve with their CSR-oriented communication. Considering that CSR-washing leads to consumers being more vigilant toward a companies' CSR communication (Ginder et al., 2021) it seems likely that the same will be true for employees and potential job applicants. Let's go back to the example of Ikea France. Despite the lauding description of the great sustainable environment and workforce, the company was condemned by a French court in June of 2021 for illegally surveilling its own employees (e.g., Arama, 2021). This conviction came only two years after Ikea France was condemned for actively trying to burry sexual harassment allegations (e.g., Leplongeon, 2019). Ikea is definitely not the only organization to be confronted with lawsuits but it is interesting to see the discrepancy between the companies' *talk* and its *walk* (Walker & Wan, 2012) and one is likely to ask what is sustainable management and how does it relate to high-quality employment relationships? To answer this question, we will briefly focus on sustainable (human resource) management before continuing with examining the employment relationship in more detail.

1.2 Sustainable Management is Everywhere but Where are the Employees?

When we started working on this chapter, we asked ourselves "What does sustainable management entail?". This seemed like a pretty straightforward question. However, it turned out to be more complicated to find a straightforward answer. A first tour of the different academic databases, using the search term "sustainable management", resulted in an impressive number of publications in the last few decades. Figure 1, which was created using the Dimensions.ai website on September 13th, 2021, clearly shows a steep uptrend in the number of articles and chapters published since 1990. Especially in the years following the 2008 financial crisis there was increased interest in the topic of sustainable management.

Although this overview gives us a brief insight into the sheer number of publications on sustainable management over the past years, it turned out to be challenging to get a grasp of the number of publications on *sustainable employment relationships*. In order to get a better idea of how often sustainability was used in relation to one's employment relationship, we started with a basic search for "sustainable management" on Ebsco and Science Direct and surprisingly enough nearly none of the publications spoke about one's employment relationship or made reference to managing people. Instead, the top results focused on diverse topics such as managing a sustainable supply chain, managing a sustainable blockchain, managing bio- or other types of waste in a sustainable way, managing ground water, implementing sustainable finance, or creating sustainable urban growth. As the employment relationship falls within the larger human resource domain, we then decided to narrow our search term by including "human resources". These changes to our search terms produced results

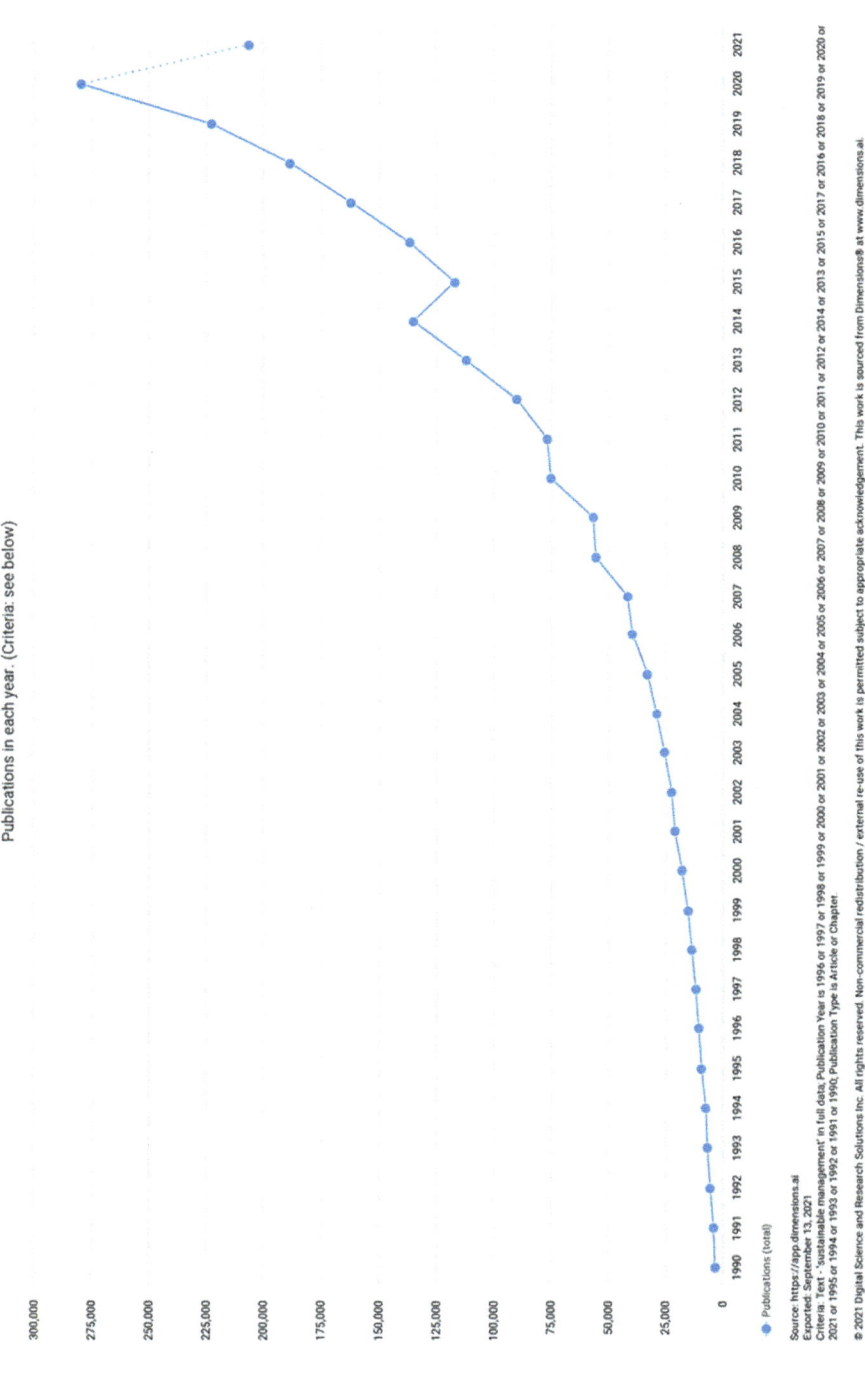

Fig. 1 Overview of published articles and chapters on sustainable management annually from 1990 to 2021

that focused on managing people, although it was still very challenging to find publications which referenced developing and maintaining a high-quality employment relationship. Although some publications made reference to employment relationships (e.g., Ehnert, 2009) these are limited to stand alone remarks and as a result we did not find any publications that specifically focused on the employment relationship within the sustainable human resource literature. The reasons for this are likely related to the observations by Järlström and colleagues (2018) who state that the field of sustainable human resources is relatively new and that it covers a wide variety of different topics. It would seem that, apart from a few brief references, the literatures on employment relationships and sustainable human resource management have not yet crossed paths in a way that produces a solid basis for understanding the individual employee and his/her employment relationship in the sustainable human resource literature.

1.3 Increasing the Focus on Employees in Sustainable Management

Next, lets focus on the employment relationship. The reader of this chapter might ask themselves why we want to examine the employment relationship in a book on sustainable management as the long-term character and the idea of safeguarding resources for the future would appear synonym to sustainability employment principles. However, our understanding of the sustainable human resource literature and practice is that human resources practices are currently fragmented. There is attention for topics such as inclusivity and gender equality (Santana & Lopez-Cabrales, 2018) but this is often treated as a standalone topic and there is no holistic approach that combines strategic goals such as general employee wellbeing, retention or organizational citizenship behavior within a structured human resource policy that aims to realize these goals. Instead, companies are increasingly focusing on short-term performance indicators (Marchington, 2015) and put stakeholder interest and profit taking before employees (Dundon & Rafferty, 2018). According to Richards (2019), sustainable human resource management is primarily employer-driven and aimed at increasing employee productivity. A recent Oxfam report (2020) on the French companies traded on the CAC40 confirmed these trends by highlighting that the past decade had seen an increase in dividend payments to shareholders instead of money being invested in more decent salaries for entry-level jobs or reducing the gender pay gap. We therefore want to present an employee-focused approach to sustainable human resource management through the lens of psychological contract research, which is a theoretical framework commonly used to study the employment relationship (Freese & Schalk, 2008). As the employment relationship is an employee's principal source of rights and benefits in the work setting (International Labour Organization, 2021) it makes sense to use the psychological contract as a starting point for working our way toward sustainable human resource practices and the eventual outcomes such as employee wellbeing and retention.

The psychological contract is comprised of a set of perceived implicit reciprocal obligations that employees and their employer have of each other in an ongoing exchange relationship (Conway & Briner, 2005; Rousseau, 1995). These obligations can be met (i.e., psychological contract fulfillment; Lee et al., 2011) but the employer can also fall short of delivering on obligations (i.e., psychological contract breach; Morrison & Robinson, 1997). The psychological contract literature generally assumes that high-quality exchange relationships are those that (1) are characterized by fulfilled obligations between both parties (Hansen & Griep, 2016) or (2) are more relational in nature (i.e., comprised of unspecified exchanges and aimed at developing and maintaining a more long-term exchange relationship, Robinson et al., 1994; Rousseau & McLean Parks, 1993). However, as we will discuss in the remainder of this chapter, the concept of high-quality psychological contracts is complex and these exchange relationships are not always accessible to every single employee.

We only found one theoretical publication (Susomrith, 2020) that explicitly focused on integrating the psychological contract in sustainable human resources and more specifically the sustainable human resources model proposed by Kramar (2014). In their publication, Susomrith (2020) considered the psychological contract to be an outcome of human resource practices. This is not surprising as Kramar (2014, p. 1076) states that the psychological contract is the outcome of human resources management practices, together with job satisfaction and engagement. However, we believe that the psychological contract is much more complex than simply an outcome of human resource practices and that said practices are actually the outcome of the organization's strategic decision on what type of exchange relationship they want to build with their employees. Hence, we position that the human resource practices are the contents or terms of the exchange relationship and that the employee perceptions regarding the quality of this relationship will influence the desired outcomes such as job satisfaction, engagement, organizational citizenship behavior, and willingness to promote the company to outsiders. Due to this seemingly oversimplification of the psychological contract in the sustainable human resource management literature thus far, we want to provide a more complete picture of what the psychological contract is and how it can be used in developing a high-quality employment relationship.

In order to provide the reader with a more complete understanding of the psychological contract in a sustainable human resource environment, the remainder of this chapter will focus on three main topics. First, we will discuss the development of the psychological contract during recruitment and socialization and will then discuss groups that are potentially left out when it comes to the creation or maintenance of said high-quality employment relationships. In this section, we also want to include macro-trends of the gig economy by including employees in entry-level jobs and self-employed individuals. Second, we will focus on the different types of psychological contract contents to have a better understanding of the areas where employer can steer toward more sustainable exchange relationships. Third, we will discuss the quality of the employment relationship, which we believe is more complex than a tale of establishing breach or fulfillment of mutual obligations between employers and their employees. In this part we will present the more nuanced reality of actual levels of fulfillment as well as the concepts of under- and over-fulfillment (i.e., delivering below or above the expected levels of obligations; Lambert et al., 2003). Finally, we will discuss what academics and practitioners can do regarding sustainable exchange relationships so that they can be studied, developed, and maintained going forward.

2 The Psychological Contract

Grounded in Social Exchange Theory (Blau, 1964), the psychological contract constitutes "an individual's beliefs regarding the terms and conditions of a reciprocal exchange agreement between that focal person and another party" (Rousseau, 1989, p. 123). As these interactions between parties continue over time, a set of perceived mutual obligations emerges (Cropanzano & Mitchell, 2005). In other words, if one party (e.g., the employer) delivers on one or more of these obligations, the other party (e.g., the employee) will respond accordingly to keep the balance in the exchange relationship. In the social exchange literature this is referred to as the reciprocity principle (Gouldner, 1960), which can have a positive or a negative orientation (Cropanzano & Mitchell, 2005). If an employee perceives that the employer has delivered on its obligations they will generally reciprocate positively. Regarding the context of this chapter, the basic premise would be that mutual fulfillment of the psychological contract would constitute a sustainable employment relationship. However, if the employer is perceived as not having upheld one or more obligations, the employee is likely to react through negative reciprocity which may have a negative downstream effect on a wide variety of emotions, attitudes, and behavior (for a meta-analysis see Bal et al., 2008; Zhao et al., 2007).

The psychological contract has been used extensively to study social exchanges at work (Kraak & Linde, 2018). Even though the term "psychological work contract" was first used by Argyris (1960, p. 97) it was not broadly used for over three decades. This all changed when Rousseau (1989, 1995) repositioned the psychological contract and provided conceptual clarity as to what it entailed (Conway & Briner, 2009). This repositioning occurred during a time in which the employment relationship was changing toward an increasingly individual accord between employers and employees (c.f., the discussion on old and new employee contracts by Kissler, 1994 or the introduction of the boundaryless career concept by Arthur & Rousseau, 1996). This increased individualism matched perfectly with Rousseau's conceptualization of the psychological contract as a collection of individual employee beliefs regarding the exchange relationship with their employer, existing "in the eye of the beholder" (Rousseau, 1989). In other words, every employee has a potentially unique psychological contract which can differ from the psychological contracts that other employees in the same organization hold with their employer (Taylor & Tekleab, 2004).

2.1 High-Quality Psychological Contracts: Development and Improvement for All?

It is generally assumed that the psychological contract is formed upon organizational entry. However, the psychological contract already starts to take shape during the recruitment process (e.g., De Vos et al., 2009; Rousseau, 1995), akin to being influenced by prior life and/or work experiences (Conway & Briner, 2009; Rousseau & Schalk, 2000). This implies that newcomers start forming a preliminary psychological contract with specific expectations about high-quality employment relationships when the company uses a CSR-oriented communication during the recruitment and selection phase. This preliminary psychological contract is further developed during the socialization process during which new employees will compare actual experiences with their initial beliefs (Rousseau, 2001). It is therefore important that the company walks their CSR-talk if they want to develop and maintain high-quality psychological contracts. It is however noteworthy that not every employee has access to such a high-quality psychological contract. This stands in stark contrast to the sustainable human resource message of most companies these days and merits further investigation. In this paragraph we would like to discuss three specific groups that have regularly been in the press over the last few years: (1) employees in entry-level jobs, (2) people who can only work for a company if they are self-employed, and (3) employees from different stigmatized groups.

The first category we want to discuss are employees in entry-level jobs. Even if companies are communicating about sustainable human resource management, this does not imply that every employee has access to a high-quality psychological contract. There have been many reports in the press regarding the exploitative nature of employment relationships for entry-level jobs such as people working at Amazon warehouses across the United States and Europe (e.g., Le Figaro & Agence France-Presse, 2019). Other examples come from the international human resource literature where easily replaced employees in production plants of foreign subsidiaries are confronted with a hard approach to human resources (Fombrun et al., 1984) focusing on managing resources, whereas employees with a rare skill set in the research and development department in the parent country enjoy a soft approach of human resources (Beer et al., 1984) that focuses on retention and commitment by keeping them happy. Although there are many people in entry-level jobs, this seems to be somewhat of a forgotten group and as a result the psychological contract literature typically focusses on professional and white-collar samples (O'Leary-Kelly et al., 2014). However, it is acknowledged that entry-level workers have little bargaining power or protection, limited chances on the labor market, and few possibilities to improve their psychological contract, resulting in beliefs of exploitation (Tomprou & Bankins, 2019). For instance, Kraak and Altman (2016) reported on self-initiated expatriates, working as waiting and kitchen staff in the French hospitality industry, who explained that their employer was legally bound to offer them permanent contracts but would purposefully breach the psychological contract until the employees would reach a breaking point after which they would either accept the exploitative character of the employment relationship or would choose to leave the company voluntarily.

The second group of people who do not have access to high-quality psychological contracts is the ever-increasing group of the self-employed or freelancers. It is well-known that Uber drivers, couriers, and the people delivering our meals are self-employed. However, the trend to hire self-employed people goes much further than that. Professionals in construction, hospitality and even teachers are increasingly hired as self-employed external staff. As a result, the volume of individuals who are no longer benefitting from the framework of an employment relationship has increased considerably. For instance, the number of self-employed people in France has gone up from approximately 600,000 at the end of 2010 to 1,900,000 in June 2020 (Acoss, 2021). These individuals do not have an employment relationship but are instead external parties with whom the organization has legal contracts. Individuals with these types of legal contracts do not enjoy the same psychological contract terms that employees who have a more traditional employment relationship with the same employer. A fitting example is the recent statement of a Dutch economist stating that salary inequalities between employees and freelancers have increased in the Netherlands since the beginning of the Covid-crisis because companies were only obligated to protect the employment status of those employees who had a permanent contract but not those on temporary or hollowed-out legal contracts (NOS, 2021). The self-employed or freelance status of these workers has resulted in a situation where they are sometimes working for longer periods of time at a company without having any kind of security regarding their employment. Furthermore, they do not have the same benefits and protection that permanent employees have (e.g., bonuses or retirement benefits; Lemsom, 2021). If we look at the list of indicators of exploitation, which was published by the International Labour Organization (2009), it clearly states that low salaries, no access to education or healthcare would count as exploitative employment if these individuals were employees of the same organization. There have been pushbacks against the "uberization of work" and these initiatives seem to be gaining traction. In the summer of 2021, a Californian judge declared a gig worker law unconstitutional because these self-employed workers only receive limited benefits (e.g., not eligible for compensation in the case of workplace accidents and no right to unionize) and cannot gain employee status (Conger & Browning, 2021). In the Netherlands representative bodies for both employees and employers are promoting more permanent employment contracts and have proposed several measures that would provide better protection to freelancers such as minimum wage, a minimum number of hours per contract, and a possible obligation to get health insurance (Lemsom, 2021).

The third group constitutes a collection of different employees having to deal with stigmatization and stereotypes such as people with different ethnic or religious backgrounds, members of the LGBTQ community, individuals having to deal with sexism or ageism, and people with handicaps or (mental) health issues (Follmer & Jones, 2021). This group is made up of many subgroups, each with specific challenges, which we cannot discuss in depth in this chapter. However, it is important to include these employees in this discussion because they regularly have to deal with deteriorated psychological contracts. This is particularly confronting seen as diversity and inclusion are so often mentioned by companies in their CSR-oriented communication. Some striking examples of the deteriorated and exploitative nature of psychological contracts are among others the 37.5% pay gap between white public sector employees in London and their black, Asian and other ethnic minority colleagues (Bulman & Musaddique, 2018), the 26% to 42% gap in retirement benefits between French men and women due to fewer career options available to women (Durand, 2017), older employees being refused a training or development opportunity because they are too old (Esser, 2020), or a Dutch study among 2100 participants that found that 53% had experienced ageism, 27% had experienced discrimination due to ethnic background and 18% had experienced sexism while applying for a job (Meester, 2020).

Overall, the take-home message appears to be that companies are increasingly using CSR-oriented communication when talking about how sustainable they are toward (future) employees but at the same time they are not allowing all of their workers access to a high-quality psychological contract. This misfit between their talk and walk, can have negative repercussions in terms of the companies' image for its stakeholders, including the new and existing employees who are developing or revising their, hopefully high-quality, psychological contract.

2.2 High-Quality Psychological Contracts: Features and Contents

The difficulty with measuring psychological contracts pertains to the idiosyncratic nature of the agreement, meaning that each psychological contract may include thousands of subjective contract terms (Kotter, 1973; McLean Parks et al., 1998), ultimately making it very hard to establish the quality of the relationship. Many researchers therefore focus on more common characteristics to identify specific attributes or features (Rousseau & Tijoriwala, 1998), that allow us to compare different kinds of contracts. Here we will focus on the most prototypical distinction between transactional and relational contracts that was first defined by MacNeil (1974) and that has been dominant in the psychological contract literature (Conway & Briner, 2009). The transactional feature refers to a short-term relationship that is very narrow in scope and therefore has explicit monetary terms whereas the relational feature describes a long-term relationship that is focused on maintaining the relationship, which means that exchange terms are less defined and more likely to change over time (Robinson et al., 1994; Rousseau & McLean Parks, 1993). Relational contracts are generally believed to be of higher quality because they relate positively to job satisfaction and affective commitment as well as fewer intentions to leave the organization (e.g., Kalleberg & Rognes, 2000; Raja et al., 2004). The difficulty in the use of these features lies in the dichotomy between transactional and relational features because they were originally considered as two ends of the same continuum (Rousseau, 1990), which would position any psychological contract somewhere between the two extremes. However, the literature has since used these features as separate categories (Taylor & Tekleab, 2004), making it more difficult to identify high-quality exchange terms if we already determined that the psychological contract in question is to be labeled as transactional. Consequently, we may be overlooking specific high-quality psychological contracts or miss potential gaps because we believe that we only have high-quality psychological contracts due to a "forced" two-option labeling issue.

Quality in an exchange relationship does not have to be limited to transactional or relational features. In these times of sustainability, individuals might define psychological contract quality in terms of broader terms that they find important and that serve broader goals. Such a psychological contract is referred to as an ideological contract, which was introduced and defined by Thompson and Bunderson (2003, p. 574) as "credible commitments to pursue a valued cause or principle (not limited to self-interest) that are implicitly exchanged at the nexus of the individual-organization relationship". Ideological psychological contracts capture the shared mutual agreements between employees and their organization that are built on a set of shared values, mission, and/or purpose the organization is believed to strive for (Thompson & Bunderson, 2003). At the forefront of these agreements and the crux of the ideology is the goal to benefit a third party such as a client, patient, or special interest group beyond the employee–organization dyad. Employees enter into these ideological psychological contracts with the understanding that the organization is committed to, and can legitimately support, activities that support the ideology (e.g., ensuring enough resources/supplies in a healthcare facility, developing policies and procedures that support environmentally-friendly behavior) and in return, employees contribute their time, skills, and energy to carry out activities related to the cause (e.g., providing high-quality patient care, recycling in their office). These ideological psychological contracts are unique from other transactional and relational psychological contract types in that the interaction and negotiation of obligations is focused on a larger shared ideology that is culturally or socially understood rather than a focus on the individual employee–organization interaction (Thompson & Bunderson, 2003). Key to the negotiation of these socially and culturally shared ideologies is the strong pro-social orientation often associated with ideological psychological contracts, aimed at benefitting a target external to the core operations of the organization.

Apart from assessing more general characteristics of the psychological contract that is captured by features, we can also measure specific psychological contract terms through the actual contents. Some authors prefer measuring specific contents as the psychological contract cannot be completely captured in a single dimension such as transactional or relational (Freese & Schalk, 2008). Content-related topics are quite popular and represent a volume in the psychological contract literature that is only surpassed by research on fulfillment and breach (Conway & Briner, 2005). The scope of this chapter is limited to identifying elements that are related to high-quality psychological contracts, which means that we cannot present all the content-related frameworks that exist in the literature. Although each framework can include unique exchange terms, we believe that they broadly operate in the same way, allowing for the use of a single framework[1] as an example here. We choose this particular framework—The Tilburg Psychological Contract Questionnaire or TPCQ (Freese et al., 2008)—because it represents many of the categories that we typically find in human resource practices and because it has been used in a variety of contexts (e.g., Lub et al., 2016; van Niekerk et al., 2019; Willem et al., 2010).

The TPCQ focusses on specific exchange terms in six categories: (1) terms that are related to the content of the job, (2) terms that provide various career development opportunities, (3) terms that help chart the social atmosphere at work, (4) terms that focus on the different organizational policies, (5) terms that relate to different aspects of work-life balance, and (6) terms that look at different elements of rewards. Each category is measured by four to eight specific items or contract terms. For instance, the category work-life balance is measured by the following four items: (1) consideration of personal circumstances, (2) opportunity to schedule your own holidays, (3) working at home, and (4) adjustment of working hours to fit personal life. Due to contextual differences, it may not come as a surprise that content measures often do not generalize to all types of employment or contexts (Janssens et al., 2003). For instance, in the United States the number of paid leave days or health insurance can be part of the psychological contract whereas these specific terms will have been defined by law in a typical European country. However, content measures are useful for establishing the quality of a psychological contract in a specific context. The advantage is that any of the chosen content measures can be used as a modular framework in which the company or the researcher adds, replaces, and deletes items. The content measure thus represents the building blocks of the psychological contract and the type of employment relationship that fits with the companies' human resource policy choices. If a company wants to implement specific terms under a sustainable human resource strategy, they can do so without changing the characterization of that relationship. Employers and employees can therefore increase the quality of the psychological contract by including those elements that both parties find important.

2.3 High-Quality Psychological Contracts: A Tale Beyond Establishing Breach and Fulfillment

When studying psychological contracts, it is important to determine how its contents will be measured because this will determine the level of analytical detail. In this paragraph we demonstrate how the conceptualization and measurement of fulfillment and/or breach will directly impact the perceptions of the quality of the exchange relationship. Furthermore, the usual way of establishing fulfillment and/or breach might actually provide misleading information, potentially leading to the misguided belief that employers hold high-quality psychological contracts with their employees. The psychological contract literature typically uses specific measures that

---

[1] We refer to Conway and Briner (2005; 2009) for an overview of other frameworks.

focus on explicit contents (Conway & Briner, 2005) to establish the quality of the psychological contract and they can be direct or indirect (Freese & Schalk, 2008). Direct measures (see Kickul et al., 2002 for an example) operate as followed: first, employees are instructed to indicate which inducements they believe were promised by their employer. Next, employees are asked to indicate the levels of fulfillment for each of those promised inducements. Indirect measures (see Kraak et al., 2017 for an example) add detail regarding the promised inducements by asking employees to (1) indicate the extent to which inducements were promised before (2) asking the same thing regarding delivered inducements. Once the scores are available, the second (fulfillment) score is subtracted from the first (promised levels) score, resulting in a single score indicating the direction of the congruence between both scales. These measures are therefore referred to as *directional measures* (Lambert et al., 2003). When subtracting delivered inducements from promised inducements, a positive direction indicates a deficient from the promised levels (i.e., breach) and a negative direction indicates a surplus (i.e., fulfillment). Direct and indirect measures are widely used in the literature (Freese & Schalk, 2008), which means that many people are used to a difference score indicating the quality of the psychological contract.

However, once we look more closely into the methodological issues with difference scores (for a review see Edwards & Parry, 1993) it becomes clear that this straightforward directional score does not necessarily tell the entire story of how employees perceive the quality of their exchange relationship with their employers. The first methodological issue is that the difference score does not allow to assess the actual levels at which fulfillment of the psychological contract takes place (Cohen et al., 2010). At first glance this might not seem relevant but it actually matters greatly when focusing on social exchange processes because the level of delivered inducements matters more for employees than the difference between promised and delivered inducements (Conway & Briner, 2009). To demonstrate this important difference, imagine the following example: an employer promised a raise of one hundred Euros in exchange for a specific employee contribution and this promise was kept upon completion. Now imagine the same example but this time the amount involved was five hundred Euros. The difference score would give us the same score for both examples because the employee received exactly what was promised, whereas in reality there is a significant difference between them as the employee in the first example will receive four hundred Euros less than the employee in the second example. It is therefore important to not only look at fulfillment but also at the actual level or volume at which fulfillment occurs. Scholars looking into the effects of actual levels of fulfillment have found a negative relationship between higher fulfillment levels and turnover intentions (Kraak et al., 2018) and a positive relationship between higher fulfillment levels and employee satisfaction (Irving & Montes, 2009; Lambert et al., 2003).

The second methodological issue relates to the limited data that a difference score provides us in terms of delivering above (i.e., over-fulfillment), at (i.e., fulfillment), or below (i.e., under-fulfillment) the extent to which an inducement was promised (Kraak et al., 2018). A difference score describes a classic linear relationship between the difference score and a dependent variable. When we subtract the level of delivered inducements from the levels of promised inducements, the result is a typical breach indicator (i.e., the higher the score the more the employer fell short of fulfilling its promises). However, this indicator only provides us with information from one side of fulfillment; either we look at fulfillment levels (i.e., delivered minus promised inducements) or we focus on breach levels (i.e., promised inducements minus delivered levels). Consequently, we cannot look at both at the same time. Hence, we are overlooking the dynamic nature of the social exchange relationship in which breach and fulfillment might happen simultaneously (Bankins et al., 2020). This is problematic because under-fulfillment and over-fulfillment should be treated separately and do not represent two ends of the same continuum. To date, most studies focus exclusively on under-fulfillment as an indicator of breach, with well established negative outcomes for employee emotions, attitudes, and behaviors (for a meta-analysis see Bal et al., 2008 or Zhao et al., 2007). The danger, especially when aiming for high-quality psychological contracts lies then in the idea that over-fulfillment should always be positive. However, Lambert and colleagues (2003) reported that the effects of over-fulfillment depend on the type of obligation in question. If the contract term interferes with the employees' desires, needs, or abilities (e.g., receiving much more responsibilities than what was agreed upon, resulting in a much higher workload) it is possible that over-fulfillment will have the same negative effects as under-fulfillment.

As we have seen in this paragraph, creating high-quality psychological contracts is not always a straightforward affair. Although the rule of thumb is that fulfillment is always good, we have seen that it is not so much the fulfillment itself but rather the levels at which fulfillment occurs. At the same time, we have also seen that the assumption that only under-fulfillment should be avoided and that over-fulfillment is always a good thing does not hold up in every situation. It is therefore important for companies to be conscient of the terms that they engage in as well as how they should deliver on said terms in order to increase quality in their relationships with employees.

3 Going Forward

Sustainable human resources management is not a singular concept (Järlström et al., 2018). We believe that it is currently being treated a little bit like the CSR report that major companies include in their annual report: there are a lot of generalized statements about strategy and the company invests in a number of loosely connected projects, often in the form of specific human resource practices for a limited group of employees, that do not necessarily lead to sustainable improvements for the concerned stakeholders. Instead, recent reports and papers suggest that employers are mainly looking out for one particular group of stakeholders, namely its shareholders (Dundon & Rafferty, 2018; Oxfam, 2020; Richards, 2019). Here we focus on another specific group of stakeholders, and arguably the most important group, namely the companies' employees. When we look at the variety of human resources practices that are being implemented in reference to sustainable human resources management such as age-diversity practices (Sousa & Ramos, 2019) and diversity and inclusion initiatives (Theodorakopoulus & Budhwar, 2015) one cannot help but think that companies are trying to patch up an outdated or even dysfunctional human resource management strategy in such a way that it passes the test of time. However, this approach does not necessarily fit with a world that is increasingly aware of sustainability-related topics and this reactive way of repairing gaps is probably less effective in attracting and retaining talented employees compared to the organizations that are actually anticipating on what their (future) employees want out of sustained employment.

In this chapter we have presented the case for implementing sustainable human resources management by developing and maintaining high-quality psychological contracts. We believe that focusing on establishing a psychological contract and attributing the appropriate human resource practices that make up such a high-quality relationship, will allow companies to put in place a more holistic human resource management system where employees feel that they—and their convictions regarding sustainability—are being valued. In the same way that the psychological contract was popular to study changes toward a more individualized employment relationship in the 1990s, we believe that it can be used to create high-quality relationships during this next monumental change regarding the terms of employee and non-employee groups that are employed by organizations.

As we have discussed in this chapter, companies need to focus on three main questions once they have decided to implement high-quality employment relationships. First, they need to establish the groups that will be included when developing high-quality psychological contracts as well as anticipate the potential backlash that may arise once it becomes clear that not everybody working for the company will benefit from these high-quality exchange relationships. Second, the company will have to establish a set of exchange terms that fits with their sustainable human resource strategy for the high-quality psychological contract. These terms have to be perceived as relevant by the parties involved. Third, the company needs to be aware of the consequences of delivering below, at, or above the agreed upon levels for the exchange terms as this may turn out to be detrimental for the employee outcomes that the company is trying to realize (e.g., affective commitment, employee retention, employee wellbeing, employees promoting the company to potential future employees).

References

Acoss. (2021). Conjoncture, Les Auto-Entrepreneurs. *Direction des Statistiques, des Etudes et de la Prévision, 321*.

Arama, V. (2021, June 15). *Ikea condamnée à un million d'euros d'amende pour avoir espionné ses salariés*. *Le Point*. Retrieved from www.lepoint.fr

Argyris, C. (1960). *Understanding organizational behaviour*. Dorsey.

Arthur, M. B., & Rousseau, D. M. (1996). A career Lexicon for the 21st Century. *Academy of Management Executive, 10*(4), 28–39.

Bal, P. M., De Lange, A. H., Jansen, P. G. W., & van der Velde, M. E. G. (2008). Psychological contract breach and job attitudes: A meta-analysis of age as a moderator. *Journal of Vocational Behavior, 72*(1), 143–158. https://doi.org/10.1016/j.jvb.2007.10.005

Bankins, S., Griep, Y., & Hansen, S. D. (2020). Charting directions for a new research era: Addressing gaps and advancing scholarship in the study of psychological contracts. *European Journal of Work and Organizational Psychology, 29*(2), 159–163. https://doi.org/10.1080/135 9432X.2020.1737219

Beer, M., Spector, B., Lawrence, P., Mills, D. Q., & Walton, R. (1984). *Human Resource Management: A general manager's perspective.* Free Press.

Blau, P. M. (1964). *Exchange and power in social life.* Wiley & Sons.

Bulman, M., & Musaddique, S. (2018, March 2). Black, Asian and ethnic minority public sector workers in London paid up to 37.5% less than white colleagues. *The Independent.* http://www.independent.co.uk

Cohen, A., Nahum-Shani, I., & Doveh, E. (2010). Further insight and additional inference methods for polynomial regression applied to the analysis of congruence. *Multivariate Behavioral Research, 4*, 828–852. https://doi.org/10.1080/00273171.2010.519272

Conger, K. & Browning, K. (2021, August 23). A judge declared California's Gig worker law unconstitutional. Now what? *The New York Times.* www.nytimes.com

Conway, N., & Briner, R. B. (2005). *Understanding psychological contracts at work: A critical evaluation of theory and research.* Oxford University Press.

Conway, N., & Briner, R. B. (2009). Fifty years of psychological contract research: What do we know and what are the main challenges? In J. K. Ford (Ed.), *International review of industrial and organizational psychology* (pp. 71–130). John Wiley & Sons.

Cropanzano, R., & Mitchell, M. S. (2005). Social exchange theory: An interdisciplinary review. *Journal of Management, 31*(6), 874–900. https://doi.org/10.1177/0149206305279602

De Vos, A., De Stobbeleir, K., & Meganck, A. (2009). The relationship between careerrelated antecedents and graduates' anticipatory psychological contracts. *Journal of Business & Psychology, 24*, 289–298. https://doi.org/10.1007/s10869-009-9107-3

Dögl, C., & Holtbrügge, D. (2014). Corporate environmental responsibility, employer reputation and employee commitment: An empirical study in developed and emerging economies. The *International Journal of Human Resource Management, 25*(12), 1739–1762. https://doi.org/10.1080/09585192.2013.859164

Dundon, T., & Rafferty, A. (2018). The (potential) demise of HRM? *Human Resource Management Journal, 28*, 377–391. https://doi.org/10.1111/1748-8583.12195

Durand, A. -A. (2017, March 7). Les inégalités femmes-hommes en 12 chiffres et 6 graphiques. *Le monde.* www.lemonde.fr

Edwards, J. R., & Parry, M. E. (1993). On the use of polynomial regression equations as an alternative to difference scores in organizational research. *Academy of Management Journal, 36*, 1577–1613. https://doi.org/10.5465/256822

Ehnert, I. (2009). *Sustainable Human Resource Management: A conceptual and exploratory analysis from a paradox perspective.* Physica-Verlag.

Esser, Y. (2020, August 4). Groot deel werknemers krijgt vroeg of laat te maken met leeftijdsdiscriminatie. Dit kun je zelf doen. *De Volkskrant.* www.volkskrant.nl

Follmer, K. B., & Jones, K. S. (2021). Navigating depression at work: Identity management strategies along the disclosure continuum. *Group & Organization Management.* https://doi.org/105960112 11002010

Fombrun, C. J., Tichy, N. M., & Devanna, M. A. (1984). *Strategic Human Resource Management.* Wiley.

Freese, C., & Schalk, R. (2008). How to measure the psychological contract? A critical criteriabased review of measures. *South African Journal of Psychology, 38*, 269–286. https://doi.org/10.1177%2F008124630803800202

Freese, C., Schalk, R., & Croon, M. A. (2008). De Tilburgse psychologisch contract vragenlijst. *Gedrag En Organisatie, 3*, 278–294.

Ginder, W., Kwon, W.-S., & Byun, S.-E. (2021). Effects of internal-external congruence-based CSR positioning: An attribution theory approach. *Journal of Business Ethics, 169*, 355–369. https://doi.org/10.1007/s10551-019-04282-w

Gouldner, A. W. (1960). The norm of reciprocity: A preliminary statement. *American Sociological Review, 25*, 161–178.

Griep, Y., Cooper, C., Robinson, S., Rousseau, D., Hansen, S. D., Tomprou, M., Conway, N., Briner, R., Jacqueline, A. M., Sharpio, C., Horgon ,R., Lub, X., De jong, J., Kraak, J. M., O'Donohoue, W., Jones, S. K., Vantilborgh, T., Yang, Y., Casser, V., Akkermans, J., Jepsen, D., Woodrow, C., De Jong, S., Sherman, U., Bezzina, Erdem, C., ... Achnak, A. et al. (2019). Psychological contracts: Back to the future. In Y. Griep & C. Cooper (Eds.), *The handbook on psychological contract research* (pp. 397–414). Edward Elgar Publishing.

Hansen, S. D., & Griep, Y. (2016). Psychological contracts. In J. Meyer (Ed.), *Handbook of employee commitment* (pp. 119–135). Edward Elgar Publishers, Inc. https://doi.org/10.4337/9781784711740.00019

International Labour Organization. (2009). *Details of indicators for labour exploitation, WCMS_105035.* ILO.

International Labour Organization. (2021). *Addendum à l'Étude d'ensemble: promouvoir l'emploi et le travail décent dans un monde en mutation, CEACR/XCI/2020/4.* Bureau international du Travail.

Irving, P. G., & Montes, S. D. (2009). Met expectations: The effects of expected and delivered inducements on employee satisfaction. *Journal of Occupational and Organizational Psychology, 82*, 431–451. https://doi.org/10.1348/096317908X312650

Janssens, M., Sels, L., & Van Den Brande, I. (2003). Multiple types of psychological contracts: A six-cluster solution. *Human Relations, 56*(11), 1349–1378. https://doi.org/10.1177%2F00187267035611004

Järlström, M., Saru, E., & Vanhala, S. (2018). Sustainable human resource management with Salience of stakeholders: A top management perspective. *Journal of Business Ethics, 152*, 703–724. https://doi.org/10.1007/s10551-016-3310-8

Kalleberg, A., & L., & Rognes, J., R. (2000). Employment relations in Norway: Some dimensions and correlates. *Journal of Organizational Behavior, 21*(3), 315–335. https://doi.org/10.1002/(SICI)1099-1379(200005)21:3%3C315::AID-JOB23%3E3.0.CO;2-1

Kickul, J., Lester, S. W., & Finkl, J. (2002). Promise breaking during radical organizational change: Do justice interventions make a difference? *Journal of Organizational Behavior, 23*, 469–488. https://doi.org/10.1002/job.151

Kissler, G. D. (1994). The new employment contract. *Human Resource Management, 33*(3), 335–352. https://doi.org/10.1002/hrm.3930330304

Kotter, J., & P. (1973). The psychological contract: Managing the joining-up process. *California Management Review, 15*(3), 91–99. https://doi.org/10.2307/41164442

Kraak, J. M., & Altman, Y. (2016). Psychological contracts of SIEs: The role of language proficiency and organizational embeddedness. *Academy of Management Proceedings, 11642.* https://doi.org/10.5465/ambpp.2016.11642abstract

Kraak, J. M., & Linde, B. J. (2019). The usefulness of the psychological contract in the 21st century. In Y. Griep & C. Cooper (Eds.), *The handbook on psychological contract research* (pp. 101–121). Edward Elgar Publishing.

Kraak, J. M., Lunardo, R., Herrbach, O., & Durrieu, F. (2017). Promises to employees matter, self-identity too: Effects of psychological contract breach and older worker identity on violation and turnover intentions. *Journal of Business Research, 70*, 108–117. https://doi.org/10.1016/j.jbusres.2016.06.015

Kraak, J. M., Russo, M., & Jiménez, A. (2018). Work-life balance psychological contract perceptions for older workers. *Personnel Review, 47*, 1194–1210. https://doi.org/10.1108/PR-10-2017-0300

Kramar, R. (2014). Beyond strategic human resource management: Is sustainable human resource management the next approach? *The International Journal of Human Resource Management, 25*(8), 1069–1089. https://doi.org/10.1080/09585192.2013.816863

Lambert, L. S., Edwards, J. R., & Cable, D. M. (2003). Breach and fulfillment of the psychological contract: A comparison of traditional and expanded views. *Personnel Psychology, 56*, 895–934. https://doi.org/10.1111/j.1744-6570.2003.tb00244.x

Le Figaro & Agence France-Presse (2019, July 22). *Les salariés d'Amazon se mobilisent à l'international contre leurs conditions de travail.* Le Figaro. www.lefigaro.fr

Lee, C., Liu, J., Rousseau, D. M., Hui, C., & Chen, Z. X. (2011). Inducements, contributions, and fulfillment in new employee psychological contracts. *Human Resource Management, 50*(2), 201–226. https://doi.org/10.1002/hrm.20415

Lemsom, M. (2021, June 6). *Werkgevers en vakbonden willen onzeker werk inperken en hopen op steun van nieuw kabinet.* Een vandaag. www.eenvandaag.avrotros.nl

Leplongeon, M. (2019, February 26). Ikea condamné pour harcèlement sexuel. *Le Point.* www.lepoint.fr

Lub, X. D., Bal, P. M., Blomme, R. J., & Schalk, R. (2016). One job, one deal ... or not: Do generations respond differently to psychological contract fulfillment? *International Journal of Human Resource Management, 27*, 653–680. https://doi.org/10.1080/09585192.2015.1035304

MacNeil, I., & R. (1974). The many futures of contracts. *Southern Californian Law Review, 47*, 691–816.

Marchington, M. (2015). Human Resource Management (HRM): Too busy looking up to see where it is going longer term? *Human Resource Management Review, 25*(2), 176–187. https://doi.org/10.1016/j.hrmr.2015.01.007

McLean Parks, J., Kidder, D. L., & Gallagher, D. G. (1998). Fitting square pegs into round holes: Mapping the domain of contingent work arrangements onto the psychological contract. *Journal of Organizational Behavior, 19*, 697–730. https://doi.org/10.1002/(SICI)1099-1379(1998)19:1+%3C697::AID-JOB974%3E3.0.CO;2-I

Meester, M. (2020). Merendeel werknemers ervaart discriminatie: 'Collega zette Surinaams accent op'. *Intermediair.* www.intermediair.nl

Morrison, E. W., & Robinson, S. L. (1997). When employees feel betrayed: A model of how psychological contract violation develops. *Academy of Management Review, 22*, 226–256. https://doi.org/10.2307/259230

NOS. (2021, September 21). Meer ongelijkheid tussen mensen met vast contract en freelancer door coronasteun. *Nederlandse Omroep Stichting.* www.nos.nl

O'Leary-Kelly, A., Henderson, K., Anand, V., & Ashforth, B. (2014). Psychological contracts in a nontraditional industry: Exploring the implications for psychological contract development. *Group and Organization Management, 39*(3), 326–360. https://doi.org/10.1177/1059601114525851

Oxfam (2020). *CAC40: Des profit sans lendemain? Inégalités, climat: Pistes pour bâtir l'entreprise du monde d'après.* Oxfam France.

Puncheva-Michelottia, P., Hudson, S., & Jin, G. (2018). Employer branding and CSR communication in online recruitment advertising. *Business Horizons, 61*(4), 643–651. https://doi.org/10.1016/j.bushor.2018.04.003

Raja, U., Johns, G., & Ntalianis, F. (2004). The impact of personality on psychological contracts. *Academy of Management Journal, 47*, 350–367. https://doi.org/10.2307/20159586

Richards, J. (2019). Putting employees at the centre of sustainable HRM: A review, map and research agenda. *Employee Relations: The International Journal.* https://doi.org/10.1108/ER-01-2019-0037

Robinson, S. L., Kraatz, M. S., & Rousseau, D. M. (1994). Changing obligations and the psychological contract: A longitudinal study. *Academy of Management Journal, 37*, 137–152. https://doi.org/10.2307/256773

Rousseau, D. M. (1989). Psychological and implied contracts in organizations. *Employee Responsibilities and Rights Journal, 2*, 121–139. https://doi.org/10.1007/BF01384942

Rousseau, D. M. (1990). New hire perceptions of their own and their employer's obligations: A study of psychological contracts. *Journal of Organizational Behavior, 11*(5), 389–400. https://doi.org/10.1002/job.4030110506

Rousseau, D. M. (1995). *Psychological contracts in organizations*. Sage Publications Inc.

Rousseau, D. M. (2001). Schema, promise and mutuality: The building blocks of the psychological contract. *Journal of Occupational and Organizational Psychology, 74*, 511–542. https://doi.org/10.1348/096317901167505

Rousseau, D. M., & McLean Parks, J. (1993). The contracts of individuals and organizations. In L. L. Cummings & B. M. Staw (Eds.), *Research in organizational behavior* (pp. 1–47). JAI Press.

Rousseau, D. M., & Schalk, R. E. (2000). *Psychological contracts in employment: Cross-national perspectives*. Sage.

Rousseau, D., & M., & Tijoriwala, S., A. (1998). Assessing psychological contracts: Issues, alternatives and measures. *Journal of Organizational Behavior, 19*, 679–695. https://doi.org/10.1002/(SICI)1099-1379(1998)19:1+%3C679::AID-JOB971%3E3.0.CO;2-N

Santana, M., & Lopez-Cabrales, A. (2018). Sustainable development and human resource management: A science mapping approach. *Corporate Social Responsibility and Environmental Management, 26*(6), 1171–1183. https://doi.org/10.1002/csr.1765

Sousa, I. C., & Ramos, S. (2019). Longer working lives and age diversity: A new challenge for HRM. *European Journal of Management Studies, 24*(1), 21–44. https://doi.org/10.5455/EJMS/288677/2019

Susomrith, P. (2020).incorporating psychological contract into the sustainable HRM model. In S. Vanka, M. B. Rao, S. Singh & M. Rao Pulaparthi (Eds.), *Sustainable Human Resource Management* (pp. 57–69). Springer Nature Singapore Pte Ltd.

Taylor, M. S., & Tekleab, A. G. (2004). Taking stock of psychological contract research: Assessing progress, addressing troublesome issues, and setting research priorities. In J. A.-M. Coyle-Shapiro, L. M. Shore, M. S. Taylor & L. E. Tetrick (Eds.), *The employment relationship—Examining psychological and contextual perspectives* (pp. 253–283). Oxford University Press.

Theodorakopoulos, N., & Budhwar, P. (2015). Guest editors' introduction: Diversity and inclusion in different work settings: Emerging patterns, challenges, and research agenda. *Human Resource Management, 54*(2), 177–197. https://doi.org/10.1002/hrm.21715

Thompson, J. A., & Bunderson, J. S. (2003). Violations of principle: Ideological currency in the psychological contract. *Academy of Management Review, 28*(4), 571–586. https://doi.org/10.5465/AMR.2003.10899381

Tomprou, M., & Bankins, S. (2019). Managing the aftermath of psychological contract violation: Employee–organizational interplay, calling, and socio-cognitive coping in vulnerable work populations. In Y. Griep & C. Cooper (Eds.), *The handbook on psychological contract research* (pp. 206–222). Edward Elgar Publishing. https://doi.org/10.4337/9781788115681.00019

Van Niekerk, J., Chrysler-Fox, P., & van Wyk, R. (2019). Psychological contract inducements and expectations conveyed to potential employees on organisations' websites. *SA Journal of Human Resource Management, 17*, 1–11. https://doi.org/10.4102/sajhrm.v17i0.1113

Walker, K., & Wan, F. (2012). The harm of symbolic actions and green-washing: Corporate actions and communications on environmental performance and their financial implications. *Journal of Business Ethics, 109*, 227–242. https://doi.org/10.1007/s10551-011-1122-4

Willem, A., de Vos, A., & Buelens, M. (2010). Comparing private and public sector employees' psychological contracts. *Public Management Review, 12*, 275–302. https://doi.org/10.1080/14719031003620323

Zhao, H., Wayne, S. J., Glibkowski, B. C., & Bravo, J. (2007). The impact of psychological contract breach on work-related outcomes: A meta-analysis. *Personnel Psychology, 60*, 647–680. https://doi.org/10.1111/j.1744-6570.2007.00087.x

Designing an Interactionist Approach for Sustainable HRM: Might the Borders and Overlaps with Internal Communication Lead to a Common End?

Daniel Roque Gomes and Maria João Santos

Abstract The purpose of this chapter is to present and to discuss existing approaches regarding Sustainable Human Resource Manangement (SHRM), and to analyze in what ways might Human Resource Manangement and internal communication structure a new component for the SHRM approach based on an interactionist perspective. The proposal presented here puts forward a three-way component model for approaching SHRM in conjunction with an explanation as to both why an interactionist perspective on SHRM may be relevant and why this research field holds the characteristics necessary to flourish within the overall HRM research agenda. We correspondingly present and justify a set of research propositions that from our perspective bring significant added value toward the existing research on this area of investigation.

Keywords Interactionist approach · Sustainable HRM · Internal communication

1 Introduction

Sustainability is hardly a subject new to the management literature (e.g., Aras & Crowther, 2009; Fisk, 2010; Starik & Kanashiro, 2013). Indeed, this has become a solid target of research dedicated to building upon the broader global challenge of developing resourceful and committed organizations heading along paths leading to overall sustainable development. Consequently, sustainability-related issues have gained the systematic attention of scholars from different management areas since at

D. R. Gomes (✉)
Polytechnic Institution of Coimbra, School of Education, R. Misericórdia Lagar dos Cortiços—S. Martinho do Bispo, 3045-093 Coimbra, Portugal
e-mail: drmgomes@esec.pt

ICNOVA, Instituto de Comunicação da NOVA, FSCH da Universidade Nova de Lisboa, Avenida de Berna, 1069-061 Lisboa, Portugal

M. J. Santos
Lisbon School of Economics and Management, SOCIUS—Research Center
in Economic and Organizational Sociology, University of Lisbon, 1649-004 Lisbon, Portugal
e-mail: mjsantos@iseg.ulisboa.pt

least the 1990s (e.g., Elkington, 1998; Gladwin et al., 1995). Especially over recent years, Human Resources Management (HRM) researchers have become more interested in understanding how to build Sustainable HRM (SHRM) practices supporting strategy, business and performance, while enhancing the achievement of organizational Corporate Sustainability (CS) goals and challenges (e.g., Ehnert et al., 2014; Pfeffer, 2010; Wikhamn, 2019). This implies viewing SHRM systems as vital components of CS development, contributing toward better defining the way organizations do their business.

While this idea seems to be attracting supporters, HRM researchers still appear to linger over the standpoint of drawing attention to just why sustainability is a valuable issue for consideration, thus revealing an unexplored avenue for research focused on explaining how to produce effective SHRM practices contributing to CS development goals. In effect, existing research seems to point to a dilemma to surpass, as on the one hand the idea of evidencing the impact of SHRM on business and on society is a whole-hearted one (e.g., Jerome, 2013; Randev & Jha, 2019; Raveenther, 2020), but on the other hand, the way how to proceed in action seems unanswered at the moment (e.g., Ehnert et al., 2019).

An issue of decisive relevance involves ascertaining whether the SHRM approach is able to deliver the "promise" of sustainable Human Resource Manangement practices aimed at deeply integrating employees into the organization's sustainability strategy. In keeping with this, we discuss here the relevance of an interactionist perspective within the scope of delivering this global SHRM aim. We also present a series of remarks on the borders and overlapping content between Internal Communication (IC) and HRM, as well as research propositions identifying means of empirically portraying the appropriateness of such proposals.

2   Is Sustainable HRM the New Research Unicorn or Just "Old Wine in New Bottles"?

The new millennium development agenda has brought about the widely accepted notion that countries worldwide must develop strong policies for the common purpose of promoting sustainable development. The current Sustainable Development Goals (SDGs) approved by the UN's General Assembly on 25th September 2015 launched the magnanimous idea of a common model for national governance shaped by the commitment toward ending poverty and ensuring economic prosperity while protecting the environment.

SHRM has been collecting support as a relevant area of contemporary research aligned agenda with the SDGs, especially, with the (a) decent work and (b) economic growth SDGs. Effectively, the HRM that literature has experienced met a shift in its research agenda, which corresponds to a trend drawing attention to the link between sustainability and HRM (e.g., De Prins et al., 2014; Diaz-Carrion et al., 2018; Kramar, 2014; Zaugg, 2009). In effect, HRM constitutes a research and management field with a relevant heritage with several decades of evidence linking it to performance (e.g., Guest, 1997; Huselid, 1995; Jackson et al., 2014). The 1980s, for instance, were very rewarding in demonstrating how management models explain the impact of people on organizations and how enhance performance outputs. Such was the case with the "Harvard Model" (Beer et al., 1984) and the "Chicago Model" (Fombrun et al., 1984). The former, sought to provide a territorial map of HRM functions, foreseeing stakeholder interests and situational factors. The second, striving to explain how HRM formed part of the business value chain, presented prescriptive means of promoting performance and related outputs. The 1990s made it clear to both researchers and practitioners that HRM required strategic planning closely related to the general strategy for enhancing the organization's performance (Brewster & Hegewisch, 1994). For instance, the Resource-Based View (RBV Theory) of organizations highlighted the role of organization competencies (Hammel & Prahalad, 1994; Prahalad & Hammel, 1990), and stimulated discussions on ways for organizations to deal with their core competencies and their relevance to promoting competitiveness (e.g. Gale, 1994; Reinbold & Breillot, 1993).

While the HRM research agenda across focused on these kinds of concerns, the concept of sustainability was gaining supporters and achieved open through the Brundtland Commission. This United Nations sub-organization was set up in 1983 with the aim of uniting countries around sustainable development concerns. In 1987, the Brundtland Commission was dissolved after releasing its report Our Common Future (also known as the Brundtland Report) (Brundtland Commission, 1987). The report identified multilateralism and the interdependence of nations as the most viable means of promoting sustainable development, placing environmental concerns at the core of political aims. This defined sustainable development as the "*development that meets the needs of the present without compromising the ability of future generations to meet their own needs*" (p. 43). Of further interest was how the report probably represents the first known effort to place human resources as a core critical value for pursuing sustainable development by recognizing HR development in the shape of poverty reduction and wealth distribution. At the time, the path for establishing a relationship between Sustainability and HRM seemed viable and promising in both research and practice.

Ehnert, Parsa, Roiper, Wagner & Muller-Camen (2016) define SHRM "as the adoption of HRM strategies and practices that enable the achievement of financial, social and ecological goals, with an impact inside and outside of the organization and over a long-term horizon" (p. 90). This deploys the distinctive characteristic of a long-term committed mindset of action, properly supported by strategic guidance, involving a pattern of HR practices seeking "to minimize the negative impacts on the natural environment and on people and communities, and acknowledges the critical enabling role of CEOs, middle and line managers, HRM professionals and employees" (Kramar, 2014, p. 1084) for the ways in which organizational development occurs. It consists of a viable way of enabling the "achievement of financial, social and ecological goals while simultaneously reproducing the HR base over a long term" (Kramar, 2014, p. 1084).

Interestingly this approach places SRHM on a double path for organizational development through HRM as, on the one hand, this provides a *means* to achieve CS objectives and, on the other hand, this represents an *end* in serving for the design of procedures and practices (Taylor et al., 2012). This double-path derives its justification the positive impact of HRM on diverse organizational contexts (e.g., Mayrhofer et al., 2019; Rothenberg et al., 2017), as well as on organization sustainability strategies (Opoku-Dakwa et al., 2018). Research into the broader aspects of sustainability and HRM covers a wide range of topics and concerns and requires solidly grounded approaches to meet the trends in research classification. A very pertinent classification typology was put forward by Ehnert et al. (2019), in which the authors propose four types of SHRM: Socially Responsible HRM (type 1); Green HRM (type 2); Triple Bottom Line HRM (type 3); Common Good HRM (type 4). According to these authors, the four types of SHRM encapsulate the existing trends taken by researchers before identifying SHRM type 4 as an underdeveloped research area.

SHRM type 1 approaches long-term oriented activities for the achieving of company CSR strategies, viewing HR as both an important component part of the strategy as well as a critical tool for implementation. This SHRM type seems to endow HR with an operational focus, with reporting responsibilities over meeting the global CSR strategy as well as ensuring that the development of HR systems and practices encapsulates CSR concerns. The literature identifies this type of SHRM as committed to minimizing "the negative impacts on business and to reduce business risks" (p. 4), with the objective being "to manage the economic risks associated with people management practices in the supply chain" (p. 4).

SHRM type 2 mainly deals with the environment. The guidance emphasizes the importance of securing deep employee involvement in the organization's environmental sustainability strategy. The literature on this SHRM type primarily focuses on enhancing employee awareness over the environmental dimension in which HR practices take place, such as valuing green hiring (valuing prospective employees with green attitudes).

In the case of SHRM type 3, authors explain its differentiation to other types of SHRM in accordance with its simultaneous focus on the "presumed economic, environmental and social purpose of HRM" (p. 4). The type delivers a broad understanding of SHRM in viewing HR functions as the means of people management guided toward people-oriented practices as well as incorporating the influences of social and ecological purposes. These three dimensions of sustainability are naturally intertwined (and thus the title of *triple bottom line*) and should work as a united strength driving HRM actions just as they should shape the ways in which HRM is developed and guided.

Finally, the existence of SHRM type 4 derives from the need to deliver business models "that address the call for a new paradigm by redefining the purpose of business in terms of common good values" (p. 5). This type of SHRM strives to establish a viable path of action for solving and delivering valid and sustainable responses for global problems. The type 4 literature serves to emphasize the inputs returned by HRM that "support business leaders and employees in contributing to ecological and social progress in the world" (p. 5), such as placing collective inter-ests above individual interests and deploying HR skills, competences and knowledge to contribute to the common good. For this purpose, the literature presents a set of examples of common good HRM, such as stimulating workplace democracy as a way of enhancing the voice of labor, employment creation as a response to job insecurity and unemployment in addition to business human rights as a response for dealing with in-work poverty and exploitative working conditions.

The SHRM research field seems appropriate for consistent long-term attention from the academic community due to the evolving scenario pointing to the severity of global climate changes (e.g., Füssel, 2007; Boegelsack et al., 2018), as well as producing HR measures aligned with technology and health care guidelines issued to cope with Covid-19 pandemic issues (e.g., Adikaram et al., 2021; Hamouche, 2021). As such, it would seem SHRM-related themes will likely become a very central topic of attention for managers and researchers as the future global agenda will surely elicit alternate ways of dealing with people, working and managing organizations. Business models more closely aligned with sustainability-related issues are thus more likely to rank as managerial priorities within the framework of producing relevant and impactful sustainable management outputs. Consequently, diverse management areas will likely be summoned to provide contributions to the cause of sustainability and the people-focused areas of any organization will certainly play a significant role.

In accordance with this mindset, it seems we need a pragmatic perspective on the best ways to develop SHRM practices and this may constitute a *unicorn* area of research, particularly in terms of the attentions of managers to HRM (the "uni-corn" reference serves to convey the potential value of this field of research to managers pursuing a comparison with Unicorn companies with their business models attributed great financial value). The consequence of applying a non-pragmatic way of dealing with these issues from the HRM perspective, especially in terms of involving employees in the sustainability strategy, may very well be a case for the expression of *old wine in new bottles*, as a way of characterizing a research area that has failed to set a new research agenda or providing new answers for practitioners. Within the aim of contributing to the discussion over how to render SHRM valuable to organi-zations, we propose an SHRM classification aligning around three components and aim to outline each facet alongside pragmatic paths of relevance to both managers and researchers, with a special focus on the interactionist component of SHRM.

2.1   The Three SHRM Components

From our perspective, the existing literature outlines three core components of SHRM: (1) Green HRM; (2) Ethics, Socially Responsibility and HRM; (3) Inter-actionist. Each SHRM component makes its own contribution to the broader goal of developing sustainable HRM strategies and practices. The first two components are recurrently referenced in the SHRM literature and display solid grounds of research development and recommendations for practitioners. However, the third component builds upon verification that several HRM-associated concepts are clearly not compatible with its traditional functions, such as enhancing employee participation or delivering good communication channels with employees. We propose that the third component is decisive to delivering the "promise" of SHRM and ensuring practices directed toward employees are able to unite and integrate them in conjunction with the sustainability goal. This third component opens the SHRM framework to the field of Internal Communication (IC) based on the functions and contributions that this specific area provides to any integrated SHRM approach built upon valuing and involving the workers in a long-term commitment to organizational sustainability. The term "interactionist" pinpoints the proposition of interactive contact points between IC and HRM within the scope of developing employee-focused stimuli and practices.

This three-component perspective of SHRM places people and output maximization at the core of concerns embedded in the view that sustainable organizations should build consistent and continuous participatory practices with their employees. Thus, we consider these three components require developing in an aggregated approach in order to deliver consistent strategies for fostering organizational sustainability through HRM. Furthermore, each component contains its own unique strengths and courses of action, enabling the building of pragmatic ways of delivering sustainability through the way people are managed and involved with their organizations and their sustainability strategies.

2.1.1   Green HRM

Green HRM encapsulates the orientating of all HR functions toward the environment and correspondingly rethinking the concepts, strategies and HR practices in environmentally friendly approaches pursuing ecological sustainability (Arulrajah & Opatha, 2016). This accounts for applying HR policies and practices designed to promote the sustainable utilization of the company's resources in the support of ecology and including the development of ecological awareness among employees (Bombiak & Marciniuk-Kluska, 2018).

This SHRM component is fully compatible with the SHRM type 2 put forward by Ehnert et al. (2019) as it plans for turning employees into green employees within the scope of achieving environmental goals and aims, and thus contributing toward environmental sustainability (Opatha & Arulrajah, 2014). The practical impact of Green HRM primarily depends on the shaping of HR practices for environmental causes and actions. Opatha and Arulrajah (2014) outline an interesting set of four categories for this goal: (1) green competences—knowledge and skills about greening; (2) green attitudes—cognitive, affective and behavioral orientations regarding green attitudes; (3) green behaviors—organizational and interpersonal citizenship behaviors; (4) green results—innovations and outcomes. According to these same authors, these four categories shape the characterization of organizations as deploying green employees as they need to possess these human requirements. As such, green HRM policies require the development of practices aimed at identifying and developing these four categories of requirements.

Ari, Karatepe, Rezaporaghdam and Avci (2021) share some of these concerns in their presentation of a conceptual model for Green HRM management within the framework of providing better directed answers about how to stimulate pro-environmental behaviors starting out from HRM. In accordance with these authors, Green HRM practices still fall far short of reaching agreement over their respective indicators with researchers proposing a diverse range of suitable HRM practices. Despite this, it nevertheless seems consensual that Green HRM should incite employees to develop concerns as regards environmental causes and collaborate with organizational goals for protecting the environment during the people managing processes in workplaces (Ari et al., 2020).

It is interesting to note how cont emporary research has been dealing with the Green HRM subject and the kind of considerations associated with how to actually produce Green HR practices. Around a decade ago, the volume of research outputs on Green HRM remained quite small as quite a young field still establishing its empirical and theoretical foundations (Jackson et al., 2011). It is also relevant to note that when performing analysis on this research topic regarding the criteria of contributions for practitioners, we encounter a promising field with prospects for sustainability-related issues. Therefore, Table 1 presents a summary of some of the contributions made for practitioners through Green HRM based on the research published in recent years:

In keeping with these recommendations for practitioners, Green HR needs building based on a set of practices underpinning the basic idea of green awareness and stimulation. Thus, HR policy should direct procedures and practices for improving environmental performance and, as such, HR practices, including recruitment, performance appraisal and training, should inspire and reward employee environmental concerns (e.g., Mandip, 2012; Rasmus, 2002).

Regarding particular associations between Green HRM and employee outcomes, existing research now proposes some relevant positions alongside empirical evidence sustaining positive and meaningful effects between Green HRM and work-related outputs, such as working engagement, job satisfaction (Ari et. al., 2020), commitment (Huo et al., 2020), environmental reputation (Malik et al., 2020), extra-role performance (Zhang et al., 2019) and organizational attractiveness (Chaundhary, 2020). Accordingly, promising future research on SHRM should follow this path associating Green HRM with relevant work-related outputs.

2.1.2 Ethics, Corporate Social Responsibility and HRM

In turn, Ethics, Corporate Social Responsibility (CSR) and HRM involve the applications of ethical stances to HRM and delivering viable paths to enhancing organizational social performances within wide CSR strategies. According to Greenwood and Freeman (2011), the debate in HRM ethics breaks down into two major fields: the micro-level (guided toward analyzing and accessing whether and how HR practices

Table 1 Summary of practitioner recommendations for developing Green HR

Authors	Recommendations for practitioners—stimulate sustainability with green HRM
Wehrmeyer (1996)	Recruitment to support environmental management Recruitment practices oriented toward stimulating newcomers with the company's culture and values
Ramus (2002)	Develop incentive management systems for rewarding eco-friendly internal initiatives Practices aimed at recognizing eco-friendly actions
Philips (2007)	Applicant attraction based on valuing the employer as a green employer Talent attraction based on the organization's green reputation
Stringer (2009)	Relevance of developing an applicant attraction strategy based on sharing the company's environmental concerns Intention to apply for job vacancies influenced by the company's environmental management
Mandip (2012)	Preparation of Green HR to elucidate HR procedures Recruitment, training and job performance evaluation as critical aspects of Green HR
Opatha et al. (2014)	Green competencies as a component in job specifications Evaluating job performance according to green-related criteria
Obaid and Alias (2015)	Green staffing will likely decrease negative the environmental effects of organizations Green training and development as key dimension for improving environmental performance
Trivedi (2015)	HR policy should replicate and inspire the HR goals Address actions regarding societal inclinations
Young et al. (2019)	Green recruitment, green training, green performance management, green teamwork and green rewards as relevant practices for stimulating green HRM
Pham et al. (2020)	Job description analysis, performance management, organizational culture and green health and safety are relevant indicators for green HRM
Ari et al., (2020)	Green selective staffing; Green training; Green empowerment; Green rewards; Green career opportunities; Green teamwork; Green work-life balance; Green decision making.

are ethical) and the macro-level (striving to understand whether the HR orientation system is ethically well-designed). As regards CSR and HRM, its main research direction has focused on the strategic HR position for CSR initiatives and the capacity to leverage CSR strategies to supply advantages for businesses, employees, society and the environment, identifying HR as a management role crucial to the success of CSR plans (InYang, et al., 2011). In addition, researchers also approach evaluating the impact of organization CSR on several individual outputs, such as commitment, job satisfaction and employer attractiveness (e.g., Chatzopoulou et al., 2021). This form of SHRM encapsulates Ehnert et al. (2019) SHRM types 1, 3 and 4.

Treviño et al. (1999) adopt an interesting approach to the subject of ethics and HRM in putting forward a perspective considering two distinct intervention strategies to enhance the ethical commitment of organizations: (1) HRM strategy—reinforcing individual employee values in order to ensure integrity and transparency; (2) Compliance-based strategy—making several tools available to employees designed to ensure ethical formalization for control and conformity. Furthermore, Treviño et al. (1999) report how HRM strategies serve the interest of elevating the ethical commitment prevailing within organizations, with leaders and managers playing key roles to this end. When the boards and management of organizations display strong ethical cultures, they tend to reinforce integrity within the organization and, as such, the roles played by leaders are critical to any upscaling of HRM policies and procedures. Regarding the second intervention strategy, the authors point out the relevance of compliance programs for reinforcing the role of HR managers and ensuring compliance with ethical standards. Correspondingly, the application of internal regulatory instruments represents a key factor for managing staff and achieving the integrity standards in effect.

Ethics and HRM research attracted significant interest over the past two decades, following the various corporate and financial scandals impacting on both society and workers, including Enron and Worldcom, leading for calls to raise ethical standards for the behaviors of leaders (e.g., Knights & O'Leary, 2005). Within these scenarios, several researchers have debated the role of HRM in preventing these kinds of outcomes for organizations. Thus, human resources managers emerge as ethical agents, ensuring organizations remain aligned with their strategy, mission and vision without causing self-harm due to the lack of ethical decision-making processes (e.g., Beatty et al., 2003; Wooten, 2001). As regards CSR and HRM, this area often draws on ethics and has now generated a field of research covering a wide array of concerns regarding the ethics and CSR consequences for HRM practices and human behavior in workplace contexts. By definition, CSR refers to the degree in which organizations embrace ethically guided sustainable concerns regarding the way they choose to relate with the diverse stakeholders, business and society in social, economic and environmental dimensions of action (Aguinis, 2011; European Commission, 2001).

The interests of CSR and HRM researchers and practitioners relate to the very relevant and interesting outputs associated with CSR and that strongly interrelate with organizational aspects of human performance. In addition, there are empirical findings linking CSR to organizational outputs including organizational attractiveness (e.g., Albinger & Freeman, 2000), motivation (Kunz, 2020), employee satisfaction (Barakat et al., 2016), employee commitment (De Silva & De Silva, 2021) or employee well-being (Kim et al., 2018). Accordingly, due to its multidimensionality, CSR comprises a wide range of activities and courses of action in which HRM is embedded in a sustainable path of action. In effect, this SHRM component builds upon the idea that HRM practices may be enabled by a set of criteria provided by ethics and CSR orientations ensuring HRM practices bond with the CSR dimensions spanning the social, environmental and economic orientations.

In terms of the association between CSR and HRM practices, researchers have more recently pointed out how organizations are increasingly seeking to enhance their business models and performance through CSR and deploying HRM as a viable path to achieve this. The main assumption seems to maintain this field serves as a core driver for the implementation of sustainability and CSR initiatives by organizations (e.g., Manroop et al., 2014). Corporate Social Responsibility Human Resources Management (CSRHRM) enhances the existing working conditions in organizations by influencing employee conduct through the CSR pillars in a long-term strategic organizational orientation (e.g., Shen & Zhu, 2011). In keeping with these theoretical orientations, CSRHRM has been positively associated with talent retention (e.g., Rawshdeh et al., 2018), employee well-being (Celma et al., 2018), work meaningfulness and job strain (Luu, 2021), avoidance of employee intentions to quit (Subhash & Gahlawat, 2015) and work attitudes (Kundu & Gahlawat, 2015).

2.1.3   The interactionist SHRM perspective

The interactionist component of SHRM derives from the added-value the Internal Communication (IC) area brings to the discussion on the bridges and overlaps with HRM, and especially with SHRM. The IC area traces its roots back to Communication Theory and approaches the communication activities and strategies produced within the scope of involving and informing the organization's internal publics (the staff), mostly based on internal "communication processes through which an organization creates, manages and develops itself" (Almeida, 2003, p. 24). Therefore, there is remarkable importance to appraising the contributions IC may make toward HRM development policies from a holistic perspective embedded within a sustainable commitment. The complementarity and interaction between IC and HRM may be tremendous, especially if attaining the shared mindset of a long-term strategic framework as might be the case for long-term commitments to sustainability. This SHRM component thus encapsulates the core idea of surrounding the organization's internal publics with wider assets that reach beyond the traditional by the book functions and delivery expectations of HRM. Correspondingly, we believe there is strong relevance to widening the framework of SHRM activities to include IC and thereby build interactive contact points between IC and HRM to obtain the prospects contiguous to SHRM as IC gains recognition as one of the foundations of modern organizations (Men & Stacks, 2014). This mainly arises for two reasons, on the one hand, IC is a powerful instrument for developing bonds between employees and their organizations, and on the other hand, this provides an instrument for the strategic construction of organizational management (Neto & Cruz, 2017) as in the case of sustainability.

According to Men (2014), IC constitutes a critical process for sharing information within organizations and consequently creating bonds and relationships between individuals and organizations, helping to build organizational culture and value sharing. The concept of IC value creation is very well described by Verghese (2017), arguing that IC enables (1) educating employees about the organization's culture and values,

(2) involving employees in the business, allows (3) aligning employee actions with customer needs, contributing to (4) integrating new employees, and stimulating the (5) sharing of relevant internal information, and supporting (6) leadership processes. In effect, IC is able to deliver exclusive contributions to organizations based on the development of practices aimed at internal publics. According to Myers and Myers (1982), IC fulfils three primary functions: (1) the coordination of production activities; (2) the integration of newly-hired workers; and (3) supporting innovation. Taking a similar line of reasoning, Brault (1992) lists seven main IC functions for organizations: (1) informative; (2) integrative; (3) retroactive; (4) signalling; (5) behavioral; (6) organizational change promoter; and (7) image management. According to Westphalen (1998), IC undertakes three main global functions of strong importance for organizations: (1) information exposition; (2) information transmission; and (3) including the workers in the organization's reality and purposes. These IC functions and purposes become strategic whenever developing or shaping any organization toward the different competitive agendas of its life cycle. These functions are clearly central to developing any strong and holistic sustainability strategy for HR, and are desirably performed through interactions between IC and a HRM sustainability strategy.

Ćorić, Vokic and Vercic (2020) make a very interesting contribution to this particular issue in detailing how IC includes all the communication taking place within an organization and represents the sharing of ideas, information, attitudes and emotions between the people of that organization, often with the intention of modifying behavior in order to meet organizational objectives. Accordingly, IC holds influence over designing prospective paths for an organization as well as the commitments of its internal publics to the ends required by the management's strategic orientations, as in the case of sustainability. As such, the IC impact on strategy is essential in keeping with its success in strengthening relationships between the organization and its employees by focusing on their mutual involvement (Brandão, 2018).

The IC strategy design may target a wide range of organizational contributions linked to production and coordination activities, integration, image management and organizational change and innovation (Gomes et al., 2014). Likewise, the design may seek to trigger information sharing inside the organization and further strong and trusting relationships between employees and organizations as well as between managers and organizations. Whenever developed according to a symmetrical communication perspective, IC will stimulate openness and reciprocity around values, practices and culture, with negotiation and openness between organizations and employees. The symmetric communication model (Grunig, 1989) may be implemented by organizations seeking mutual agreements and proximity between their component parts, within the goal of promoting and accepting ideas and behaviors on a basis of mutual understanding and acceptance (Yue et al., 2020). All of these activities and orientations are essential for worker-focused strategic and sustainable projects and encounter a greater likelihood of success when intertwined with HRM policies fostering interactions between HRM practices and IC.

From a pragmatic perspective, the interactionist SHRM component needs a two-way alignment of its orientations: (1) HRM and IC interactive policies in order to identify entangled procedures for valuing the workers—both HRM and IC have their functions and practices established by internal organizational procedures. These procedures should be developed and angled toward collaboration between the sectors of activity and seeking the scope for interaction. This joint collaboration should take into account sustainability concerns, notably, worker-focused issues. For instance, when acting on recruitment and selection, procedures should assure the organization will value applicants with strong sustainability concerns in their profiles and there are internal communication efforts guaranteeing that existing workers are aware of vacant job opportunities through strong communication practices as a means of stimulating internal sustainability through opportunity sharing. These communication efforts should convey solid and symmetrical message content on sustainability concerns; (2) HRM and IC interactive policies align with organization strategic and sustainability orientations—HRM and IC orientations should share common ground and comply with the organization's global strategic aims as well as the sustainability orientations in terms of formulating such strategies. Thus, as both areas will contribute to long-term organizational goals, they should be dealt from the perspective of maximizing joint contributions for common grounds. For instance, when the organization's global plan foresees maximizing innovative competences for sustainability, the HR strategy should direct its practices toward this aim in conjunction with the prevailing IC practices. The means of drafting and developing the annual internal communication plan should incorporate the strategic HR planning for a similar time schedule in conjunction with a global alignment toward maximizing sustainability-related innovative competences.

2.2 Bridging to SHRM: in View of the Borders and the Overlaps

Traditional HRM functions and practices seek to surround the people in organizations with sets of activities able to capture the organization's best interests regarding the skills and competences of their employees and thus "unite and guide the employees in line with the business strategies" (Hsu & Wang, 2012). Today's vision regarding this management field identifies the human factor as indispensable to generating competitive advantages, achieved through deploying solid practices directing employee skills, capacities and knowledge (Hamadamin & Atan, 2019).

To deliver value for organizational goals and competitiveness, HR practices seek to capture human value throughout a worker's life cycle in the organization in conjunction with a mindset of absorbing skills and competences. Recruitment, integration, performance appraisal, rewards and benefits are examples of some of these practices aimed at generating value through managing human capabilities within the framework of the organizational strategy. Internal communication, on the other hand, applies a different mindset of actions in which the value proposal starts with viewing the workers as internal publics, who need information and inclusion to participate in the organization's activities. This internal publics perspective brings about a set of practices quite different to those foreseen by HRM, such as internal communication planning, developing speak up practices and information resources, targeting information, updating the profiles of worker information needs.

Despite the differences in approach between HRM and IC, the traditional functions of HRM and IC share a common bond: the workers. Both fields aim to develop resources favoring worker activities within their production processes. This common bond may surely be enriched when embodied in a two-way alignment of orientations between the interactionist SHRM component: HRM and IC interactive policies striving to leverage entangled procedures that value workers in conjunction with HRM and IC interactive policies aligned with the organizational strategic orientations and strategic sustainability orientation. Accordingly, when sharing this mindset, organizational researchers and practitioners may engage in a strong and fertile area of activity. In effect, the SHRM area of research resulting from this reasoning may emerge as a unicorn area of research given the renowned interest of academia in HRM-related sustainability issues and the prospect of sustainability issues remaining predominant for decades to come. On the other hand, practitioners are currently eager for know-how about how to act in an era when organizations are highly pressured to deliver management processes designed according to sustainable premises.

For researchers, and according to our view of SHRM, this area contains very pertinent and distinctive contributions to existing knowledge, notably for understanding the interactions between HRM, IC and sustainability. In practice, this research field displays characteristics that underpin the opinion that it represents a strong and prospective area for forthcoming years, such as the potential for added-value, innovativeness and the ability to aggregate contributions from several subjects involved in people management related issues. Hence, this hardly amounts to a case of applying the well-known "old wine in new bottles" expression. In accordance with our perspective, this field enables the outlining of research propositions potentially able to enrich the existing literature through returning new and insightful knowledge as well as relevant recommendations for practitioners.

In our perspective, the existing SHRM research is at a stage when research into two major facets is desirable: (1) the consequences of SHRM practices on worker attitudes and behaviors; (2) the processes interfering with the impacts of SHRM practices. Indeed, despite the existence of strong contributions to this issue of the SHRM consequences (e.g., Ari et. al., 2020; Huo et al., 2020) for the alignment of Green HRM or Ethics, CSR and HRM, we found no record of research into the interactive abilities of HRM and IC for sustainability and the consequences for worker attitudes and behaviors. Likewise, the literature does not seem to contain any research findings on the distal processes explaining the impacts of the combined influence of HRM and IC for sustainability over SHRM practices on worker attitudes and behaviors. Following this reasoning, the first research proposition we would like to outline is:

RP1: to understand the influence of the interactive effect between HRM and IC for the sustainability of individual worker performances.

As explained, the combined influences of HRM and IC on sustainability are still unknown in terms of the consequences for worker performance. Clarifying this research proposition would be of great value to existing research as it would bring clarity to the combined influence of HRM and IC for performance sustainability. For practitioners, this would represent a valid path toward incentivizing the adoption of interactive HRM and ICl practices and understanding how to best develop these practices and stimulate individual performance levels.

A second research proposition we would outline here is:

RP2: to understand the influence of the interactive effects between HRM and IC for sustainability in the workplace attitudes of workers.

The combined influence of HRM and IC over sustainability is also unknown as regards the consequences for worker attitudes, such as commitment, employer attractiveness or perceived organizational support. This research proposition might generate a compelling research trend in this field with relevant contents for practitioners and researchers as a prospective viable way of stimulating bonds and quality in workplace relationships between employers-employees.

A third and final research proposition involves the following:

RP3: to identify the mediating effect of worker commitment and perceived organizational support in the relationship between the interactive effects between HRM and IC for sustainability on individual worker performances and attitudes.

Our third research proposition proposes that the interactive influences of HRM and IC over sustainability may activate a distal process, combining moderation and mediation effects, explaining worker performances and workplace attitudes. In practice, there is as yet no research available that focuses on discovering in what way the combined influence of HRM and IC holds the ability to activate workplace attitudes and behaviors. Clarifying this research proposition would also bring value for practitioners as this might return a viable path for stimulating distal processes for daily management capable of strengthening the relationship quality between workers and their organizations.

3   Concluding Remarks

Since the advent of this new century, sustainability-related issues have gained central importance throughout society, branding the discourses of politicians, entrepreneurs, practitioners and citizens. Following recognition of this centrality, researchers have become interested in understanding the ways sustainability fits into other fields beyond environmental considerations. The interest sustainability has attracted among HRM researchers derives from this perspective (e.g., Wikhamn, 2019).

Regarding the ways HRM researchers have approached the subject of sustainability, the question emerging is: does SHRM hold the characteristics of becoming a *research unicorn* for researchers in this field or are the ideas surrounding SHRM just *old wine in new bottles*? Answering this question implied reviewing the ways the literature considers the main SHRM-associated results alongside efforts to identify the virtues and limits of existing research. In accordance with this purpose, we here propose a three-component SHRM model in order to evaluate the knowledge existing on this subject: (1) green HRM; (2) ethics, social responsibility and HRM; (3) interactionist.

Our proposal outlines how existing research has primarily focused on the alignment between the first two component and how the interactionist component might bring about significant added value to existing research based on the contents of HRM functions and the SHRM demanded degree of commitment and actions. Therefore, the interactionist component generates the novel contribution of considering how HRM and IC require joint development and interaction due to the interrelated IC and HRM functions and the benefits of this alignment from the sustainability perspective.

The research propositions we have presented potentially deliver an interesting perspective regarding empirical evidence supporting the extent to which the interactions between HRM and IC produce relevant consequences for organizational activities, most notably those related with employee attitudes and behavior.

Acknowledgements This work was supported by FCT, I.P., the Portuguese national funding agency for science, research and technology, under the Project UIDB/04521/2020.

References

Adikaram, A., Pryiankara, H., & Naotunna, N. (2021). Navigating the crises of COVID-19: Human resource professionals battle against the pandemic. *South Asian Journal of Human Resource Manangement, 1*, 1–27.

Aguinis, H, (2011). Organizational responsibility: Doing good and doing well. In S. Zedeck (Ed.), *APA handbook of industrial and organizational psychology* (Vol. 3, pp. 855–879).

Albinger, H. S., & Freeman, S. J. (2000). Corporate social performance and attractiveness as an employer to different job seeking populations. *Journal of Business Ethics, 28*, 243–253. https://doi.org/10.1023/A:1006289817941

Almeida, V. (2003). *A comunicação interna na empresa.* Área Editora.

Aras, G., & Crowther, D. (2009). *The durable corporation: Strategies for sustainable development.* Gower Applied Business Research.

Ari, E., Karatepe, O. M., Rezapouraghdam, H., & Avci, T. (2020). A conceptual model for green human resource management: Indicators differential pathways and multiple pro-environmental outcomes. *Sustainability, 12*(17), 7089. https://doi.org/10.3390/su12177089

Arulrajah, A., & Opatha, H. (2016). Analytical and theoretical perspectives on Green Human Resource Manangement: A simplified underpinning. *International Business Research, 9*(12), 153–164.

Barakat, S. R., Isabella, G., Boaventura, J. M. G., & Mazzon, J. A. (2016). The influence of corporate social responsibility on employee satisfaction. *Management Decision, 54*(9), 2325–2339. https://doi.org/10.1108/MD-05-2016-0308

Beatty, R., Ewing, J. R., & Tharp, C. (2003). HR's role in corporate governance: Present and prospective. *Human Resource Manangement, 42*, 257–269.

Beer, M., Spector, R., Lawrence, M., Quinn, E., & Walton, E. (1984). *Managing human assets: The ground breaking Harvard Business School Programs*. Free Press.

Boegelsack, N., Withey, J., O'Sullivan, G., & McMartin, D. (2018). A critical examination of the relationship between Wildfires and climate change with consideration of the human impact. *Journal of Environmental Protection, 9*, 461.

Bombiak, E., & Marciniuk-Kluska, A. (2018). Green Human Resource Manangement as a tool for the sustainable development of enterprises: Polish Young company experience. *Sustainability, 10*(1739), 1–22.

Brandão, N.G. (2018). A comunicação interna como reforço da valorização das pessoas e seus níveis de engagement nas organizações. *Revista Media & Jornalismo*. 91–102.

Brewster, C., & Hegewisch, A. (1994). *Policy and practice in European Human Resource Manangement: Evidence and analysis*. Routledge.

Brundtland Commission, B. (1987). *Report of the World Commission on environment and development: Our common future*. Oxford University Press.

Celma, D., Martinez-Garcia, E., & Raya, J. M. (2018). Socially responsible HR practices and their effects on employees' wellbeing: Empirical evidence from Catalonia, Spain. *European Research on Management and Business Economics, 24*, 82–89.

Chatzopoulou, E., Manolopoulos, D., & Agapitou, V. (2021). Corporate social responsibility and employee outcomes: Interrelations of external and internal orientations with job satisfaction and organizational commitment. *Journal of Business Ethics*. https://doi.org/10.1007/s10551-021-048 72-7

Chaundhary, R. (2020). Green Human Resource Manangement and job pursuit intention: Examining the underlying process. *Corporate Social Responsibility Environmental Management, 26*, 929–937.

Ćorić, S., Vokić, N., & Verčič, A. (2020). Does good internal communication enhance life satisfaction? *Journal of Communication Management, 24*(4), 363–376. https://doi.org/10.1108/JCOM-11-2019-0146.

De Prins, P., Van Beirendonck, L., De Vos, A., & Segers, J. (2014). Sustainable HRM: Bridging theory and practice through the "Respect Openness Continuity Model. *Management Review, 25*, 263–284.

De Silva, K. M., & De Silva Lokuwaduge, C. S. (2021). Impact of corporate social responsibility practices on employee commitment. *Social Responsibility Journal, 17*(1), 1–14. https://doi.org/10.1108/SRJ-01-2019-0027

Diaz-Carrion, R., Lopez-Fernandez, M., & Romero-Fernandez, P. (2018). Evidence of diferente models of socially responsible HRM in Europe. *Business Ethics European Review, 28*, 1–18.

Ehnert I., Harry W., & Zink K. J. (2014) Sustainability and HRM. In I. Ehnert , W. Harry, K. Zink (Eds.), *Sustainability and Human Resource Manangement*. CSR, Sustainability, Ethics & Governance. Springer. https://doi.org/10.1007/978-3-642-37524-8_1

Ehnert, I., Matthews, B., & Muller-Camen, M. (2019). Common Good HRM: A paradigm shift in Sustainable HRM? *Human Resource Manangement Review, 30*. https://doi.org/10.1016/j.hrmr.2019.100705

Ehnert, I., Parsa, S., Roper, I., Wagner, M., & Muller-Camen, M. (2016). Reporting on sustainability and HRM: A comparative study of sustainability reporting practices by the world's largest companies. *The International Journal of Human Resource Management, 27*(1), 88–108.

Elkington, J. (1998). *Cannibals with forks: The triple bottom line of 21st century business*. New Society.

European Commission (EC). (2001). *Green paper. Promoting a European framework for corporate social responsibility*. http://europa.eu/rapid/press-release_DOC-01-9_en.pdf. Accessed on September 7, 2021.

Fisk, P. (2010). *People, planet, and profit: How to embrace sustainability for innovation and business growth*. Kogan Page.

Fombrun, C., Tichy, N., & Devanna, M. (1984). *Strategic Human Resource Manangement*. John Wiley and Sons Inc.

Füssel, H.-M. (2007). Vulnerability: A generally applicable conceptual framework for climate change research. *Global Environmental Changes, 17*, 155–167.

Gale, B. (1994). *Managing customer value—Creating quality and service that customers can see.* The Free Press.

Gladwin, T. N., Kennelly, J. J., & Krause, T. (1995). Shifting paradigms for sustainable development: Implications for management theory and research. *Academy of Management Review, 20*, 874–907.

Gomes, D., Asseiro, V., & Ribeiro, N. (2014). Socially responsible internal communication? Analysing the combined effect of CSR andInternal communication on employee's affective bond to organization. *International Journal of Marketing, Communication and New Media, 2*, 168–190.

Greenwood, M., & Freeman, R. (2011). Ethics and HRM. *Business and Professional Ethics Journal, 30*, 269–292. https://doi.org/10.5840/bpej2011303/413

Grunig, J.E. (1989). Symmetrical presuppositions as a framework for public relations theory. In C. Botan & V. Hazleton Jr. (Eds.), *Public Relations Theory* (pp. 17–44). Lawrence Erlbaum Associates.

Guest, D. (1997). Human Resource Manangement and performance: A review and research agenda. *The International Journal of Human Resource Manangement, 8*, 263–276.

Hamadamin, H., & Atan, T. (2019). The impact of strategic Human Resource Manangement practices on competitive advantage sustainability: The mediation of human capital development and employee commitment. *Sustainability, 11*, 1–19.

Hammel, G., & Prahalad, C. (1994). *Competing for the future.* Harvard Business School Press.

Hamouche, S. (2021). Human Resource Manangement and the COVID-19 crisis: Implications, challenges, opportunities, and future organizational directions. *Journal of Management & Organization, 1*, 1–16. https://doi.org/10.1017/jmo.2021.15

Hsu, L. C., & Wang, C. H. (2012). Clarifying the effect of intellectual capital on performance: The mediating role of dynamic capability. *British Journal of Management, 23*, 179–205.

Huo, W., Li, X., Zheng, M., Liu, Y., & Yan, J. (2020). Commitment to Human Resource Manangement of the top management team for green creativity. *Sustainability, 20*, 1008.

Huselid, M. (1995). The impact of Human Resource Manangement practices on turnover, productivity and corporate financial performance. *Academy of Management Journal, 38*, 635–670.

InYang, B., Awa, H., & Enuoh, R. (2011). CSR-HRM nexus: Defining the role engagement of the human resources professionals. *International Journal of Business and Social Science, 2*(5), 118–126.

Jackson, S. E., Renwick, W. S., Jabbour, C. J. C., & Muller-Camen, M. (2011). State-of-the-art and future directions for green Human Resource Manangement: Introduction to the special issue. *German Journal of Research in Human Resource Manangement, 25*(2), 99–116.

Jackson, S., Schuller, R., & Jiang, K. (2014). An inspirational framework for strategic Human Resource Manangement. *Academy of Management Annals, 8*, 1–56.

Jerome, N. (2013). Impact of sustainable Human Resource Manangement and organizational performance. *International Journal of Asian Social Science, 3*, 1287–1292.

Kim, H. (L), Woo, E., Uysal, M., & Kwon, N. (2018). The effects of corporate social responsibility (CSR) on employee well-being in the hospitality industry. *International Journal of Contemporary Hospitality Management, 30*(3), 1584–1600. https://doi.org/10.1108/IJCHM-03-2016-0166

Knights, D., & O'Leary, M. (2005). Reflecting on corporate scandals: The failure of ethical leadership. *Business Ethics: A European Review, 14*, 359–366. https://doi.org/10.1111/j.1467-8608.2005.00417

Kramar, R. (2014). Beyond strategic human resource management: Is sustainable human resource management the next approach? *The International Journal of Human Resource Management, 25*(8), 1069–1089.

Kundu, S., & Gahlawat, N. (2016). Effects of socially responsible HR practices on employees' work attitudes. *International Journal of Human Resources Development and Management, 16*, 140. https://doi.org/10.1504/IJHRDM.2016.078194

Kunz, J. (2020). Corporate social responsibility and employees motivation—Broadening the perspective. *Schmalenbach Bus Rev, 72*, 159–191. https://doi.org/10.1007/s41464-020-00089-9

Luu, T. T. (2021). Socially responsible human resource practices and hospitality employee outcomes. *International Journal of Contemporary Hospitality Management, 33*(3), 757–789. https://doi.org/10.1108/IJCHM-02-2020-0164

Malik, S., Cao, Y., Mughal, Y., Kundi, G., & Ramayah, T. (2020). Pathways towards sustainability in organizations: Empirical evidences on the role of green human resources management practices and green intellectual capital. *Sustainability, 12*, 28–32.

Mandip, G. (2012). Green HRM: People management commitment to environmental sustainability. *Research Journal of Recent Sciences, 1*, 244–252.

Manroop, L., Singh, P., & Ezzedeen, S. (2014). Human resource systems and ethical climates: A resource based perspective. *Human Resource Manangement, 53*(5), 795–816.

Mayrhofer, W., Gooderham, P., & Brewster, C. (2019). Context and HRM: Theory, evidence, and proposals. *International Studies of Management & Organization, 49*(4), 355–371. https://doi.org/10.1080/00208825.2019.1646486

Men, L. R., & Stacks, D. W. (2014). The effects of authentic leadership on strategic internal communication and employee-organization relationships. *Journal of Public Relations Research, 26*(4), 301–324.

Myers, M., & Myers, G. (1982). *Managing by communication: An organizational approach.* McGraw-Hill Book Company.

Neto, C., & Cruz, Sofia, A. (2017). Comunicação interna e comprometimento organizacional: O caso da autoridade para as condições do Trabalho. *Sociologia: Revista da Faculdade de Letras da Universidade do Porto, 34*, 47–72. https://doi.org/10.21747/08723419/soc34a3

Obaid, T., & Alias, R. (2015). The impact of green recruitment, green training and green learning on the firm performance: Conceptual paper. *International Journal of Applied Research, 15*, 951–953.

Opatha, H., & Arulrajah, A. (2014). Green Human Resource Manangement: Simplified general reflections. *International Business Research, 7*, 101–112.

Opoku-Dakwa, A., Chen, C. C., & Rupp, D. E. (2018). CSR initiative characteristics and employee engagement: An impact-based perspective. *Journal of Organizational Behaviour, 39*, 580–593.

Pfeffer, J. (2010). Building sustainable organizations: The human factor. *Academic Management Perspectives, 24*, 35–45.

Pham, N., Hoang, H., & Phan, P. (2020). Green Human Resource Manangement: A comprehensive review and future research agenda. *International Journal of Manpower, 41*, 845–878.

Prahalad, C., & Hamel, G. (1990). The core competence of the corporation. *Harvard Business Review, 68*, 79–91.

Ramus, C. A. (2002). Encouraging innovative environmental actions: What companies and managers must do. *Journal of World Business, 37*, 151–164. https://doi.org/10.1016/S1090-9516(02)00074-3

Randev, K., & Jha, J. (2019). Sustainable human eesource management: A literature-based introduction. *NHRD Network Journal, 212*(3), 241–252. https://doi.org/10.1177/2631454119873495

Raveenther, A. (2020). Impact of sustainable human resources management practices in organizational performance of banking sector in Trincomalee district in Sri Lanka. *International Journal of Advanced Engineering and Management Research, 5*, 57–67.

Rawshdeh, Z., Zain, Z., & Ismail, I. (2018). The influence of socially responsible-HRM practices on retaining talents. *International Journal of Engineering & Technology, 7*, 384–387.

Reinbold, M., & Breillot, J. (1993). *Gérer la Compétence das l'enterprise.* Harmattan.

Rothenberg, S., Hull, C., & Tang, Z. (2017). The impact of Human Resource Manangement on corporate social performance strengths and concerns. *Business and Society, 56*(3), 391–418.

Shen, J., & Jiuhua, ZhuC. (2011). Effects of socially responsible Human Resource Manangement on employee organizational commitment. *The International Journal of Human Resource Manangement, 22*, 3020–3035.

Starik, M., & Kanashiro, P. (2013). Toward a theory of sustainability management: Uncovering and integrating the nearly obvious. *Organization & Environment, 26*, 7–30. https://doi.org/10.1177/1086026612474958

Stringer L. (2009). *The Green workplace. Sustainable strategies that benefit employees, the environment, and the bottom line.* Macmillan.

Kundu, S. C., & Gahlawat, N. (2015). Socially responsible HR practices and employees' intention to quit: The mediating role of job satisfaction. *Human Resource Development International, 18*, 387–406. https://doi.org/10.1080/13678868.2015.1056500

Taylor, S., Osland, J., & Egri, C. (2012). Introduction to HRM's role in sustainability: Systems, strategies and practices. *Human Resource Manangement, 51*, 789–798.

Treviño, L. K., Weaver, G. R., Gibson, D. G., & Toffler, B. L. (1999). ManagingeEthics and legal compliance: What works and what hurts. *California Management Review, 41*(2), 131–151.

Trivedi, A. (2015). Green HR: Traditions and designed effort in the organizations. *International Journal of Management, Information Technology and Engineering, 12*, 29–36.

Verghese, A. K. (2017). Internal communication: Practices andimplications. *SCMS Journal of Indian Management,*103–113.

Wehrmeyer, W. (1996). *Greening people: Human resource and environmental management.* Greenleaf.

Wikhamn, W. (2019). Innovation, sustainable HRM and customer satisfaction. *International Journal of Hospitality Management, 76*, 102–110.

Wooten, K. C. (2001). Ethical dilemmas in Human Resource Manangement: An application of a multidimensional framework, a unifying taxonomy, and applicable codes. *Human Resource Manangement Review, 11*, 159–175.

Young, Y., Yusliza, M., & Fawehimni, O. (2019). Green Human Resource Manangement: A systematic literature review from 2007 to 2019. *Benchmarking International Journal, 27*, 2005–2027.

Yue, A., Men, L., & Ferguson, M. (2020). Examining the effects of internal communication and emotional culture on employees' organizational identification. *International Journal of Business Communication, 58*(2), 169–195. https://doi.org/10.1177/2329488420914066.

Zaugg, R. (2009). *Sustainable HR management: New perspectives and Empirical explanations.* Gabler.

Zhang, Y., Luo, Y., Zhang, X., & Zhao, J. (2019). How green human resources management can promote green employee behaviour in China: A technology acceptance model perspective. *Sustainability, 11*, 5408.

Recruitment and Selection Processes Influence on the Access of Immigrants to the Labor Market

Raquel Cerqueira Gonçalves and Carolina Feliciana Machado⦿

Abstract This chapter reflects the intention to explore the theme "Recruitment and Selection Processes Influence on the Access of Immigrants to the Labor Market." For this purpose, a research was carried out in the existing literature on this subject. It should be mentioned that it is relatively scarce from the perspective of the analysis of those responsible for the Recruitment and Selection processes. That is, there are only broader researches, from the perspective of the immigrant, to understand which are the most relevant issues to be addressed in this chapter and also what are the main conclusions that have been reached. In this way, the following themes of interest were addressed, namely, migration, immigration, immigration and its relationship with the labor market, having been studied, at this point, the theory of the segmented labor market. Next, topics related to diversity and the recruitment and selection processes were presented. And, finally, it was intended to examine the influence that the latter exert on immigrants' access to the labor market. For that, in this final topic, two of the theories that explain why there are biases in the immigrant selection processes are presented, respectively, the similarity-attraction theory and the social identity theory. That said, the main conclusion drawn was that recruiters and selectors can play a decisive role in the decision to hire immigrant candidates. This is due to the fact that these can be influenced by biases that may occur in Selection interviews. As a result, they see immigrants as different, from themselves and from other members of the organization, and believe that they will threaten the existing similarity within the company, and thus, immigrants may be harmed. This means that, in most cases, competences, which represent one of the most important aspects for the performance of a function, in the case of immigrants, are not valued or recognized.

Keywords Immigrants · Labor Market · Recruitment and Selection · Biases · Similarity

R. C. Gonçalves · C. F. Machado (✉)
Department of Management, School of Economics and Management, University of Minho, Braga, Portugal
e-mail: carolina@eeg.uminho.pt

C. F. Machado
Interdisciplinary Centre of Social Sciences (CICS.NOVA.UMinho), University of Minho, Braga, Portugal

© The Author(s), under exclusive license to Springer Nature Switzerland AG 2022   41
C. Machado and J. P. Davim (eds.), *Sustainability and Intelligent Management*,
Management and Industrial Engineering, https://doi.org/10.1007/978-3-030-98036-8_3

1 Introduction

Not being new phenomena, migratory processes have an increasingly significant impact on the dynamics of any country or society, as they can contribute in a very positive way to its development and transformation.

According to estimates in the Global Migration Report 2020, released by the International Organization for Migration, the number of international migrants in the year 2019 corresponded to about two hundred and seventy-two million. Such statistic thus surpasses, in an overwhelming way, some projections made for 2050 about their growth, as they pointed to around two hundred and thirty million migrants (Nações Unidas, 2019).

This high migration flow can be caused by a wide range of reasons. As an example, the conflicts or wars existing in the countries of origin, the increasingly extreme weather conditions, the lack of freedom of expression and the prevalence of unemployment and low salaries. However, there is also an increasing spontaneity in these processes, that is, individuals aspire to emigrate due to the desire to make changes in their lifestyle or goals related to their professional path (Connell & Burgess, 2009).

Thus, from the above, it can be seen that the increasing acceleration of migratory movements results in an increase in diversity in societies and, more specifically within the scope of this chapter, in the labor market. In other words, we are witnessing a workforce that is increasingly equipped with idiosyncrasies, which can constitute a competitive advantage for companies (El-Cherkeh, 2009). In this sense, the Sustainable Development Goals (SDGs) 8 and 10, respectively, "Decent Work and Economic Growth" and "Reducing Inequalities," fall perfectly within the scope of this chapter, since both converge toward inclusion, equal opportunities and for the empowerment of each and every individual, regardless of age, race, ethnicity, religion, gender, among others. It will also be in this way that organizations will be able to extract the maximum amount of creativity and innovation.

According to article 13, paragraph 2 of the Universal Declaration of Human Rights, every individual has "(…) the right to leave the country in which he finds himself, including his own, and the right to return to his own country." Even so, most of the time, these people who enter in another country than their own, aiming for a better or more stable life, have been associated with negative situations, such as crime, marginality or lack of capacities, skills and financial conditions. In short, the immigrant is seen as "a problem to be solved" (Rodrigues et al., 2013, p. 98) and can face serious situations of discrimination, prejudice and rejection when accessing the world of work. Therefore, this problematic of the Recruitment and Selection processes aimed at these specific collaborators can present itself as a challenge for the pursuit of the above, mentioned SDGs.

However, if the organizational sector cooperates and is able to adopt intelligent people management, the situation can be reversed. Therefore, it is vital that companies identify and bring together a set of coherent, favorable and efficient human resources practices for the adaptation and integration of immigrants and also for their development. Therefore, in this chapter, two of the human resources processes will be addressed, respectively, recruitment and selection, in order to later understand what their influence on the final decision of hiring immigrants will be.

Thus, the present chapter is divided into seven sections. In the first phase, issues relating to migratory phenomena and the immigration process are contextualized, these corresponding to the first two sections. Then, we intend to transport immigration to the approach under study, that is, the relationship between it and the labor market is analyzed, and for this, one of the theories that explains the migration processes is presented and, consequently, that also leads to an explanation for the fact that immigrants' skills are often underused. Having said that, in the fourth section, the theme of diversity is presented, as it is unavoidably linked to migration, insofar as the intensification of the displacement of people causes an increasingly diverse workforce. Regarding the fifth and sixth sections, the recruitment and selection processes are characterized and, still, in the topic related to selection, some of the biases that may occur in the interviews are addressed, to set the tone for the final section. Thus, finally, two theories are studied that clarify the occurrence of perceptual distortions in immigrant selection processes, namely the similarity-attraction theory and the social identity theory, and which, consequently, may hinder the realization of SDG 8 and 10.

2 Literature Review

2.1 Migration

It is not possible to talk about globalization and environmental and demographic pressures, without referring to the migratory processes that have been observed, in an accentuated way, nowadays. In the literature, it is widely accepted that the essential factor for migration is globalization (Massey et al., 1999). This poses a series of new challenges to societies that can be converted into changing people's jobs, moving them to other locations (which may or may not be a consequence of the above), and/or changing ways of thinking and acting. In other words, all of this could lead to the transformation of communities (Milanovic, 1999).

If we look at the definition of the International Organization for Migration (2009), migration is established as the process of crossing, by an individual or group, of their country or state, whatever its causes. According to Martine (2005), the migrant's objective is no longer limited to moving to the capital or to a larger city in his/her country. On the contrary, his/her horizon is the world. However, it is necessary to pay attention that these transitions entail both psychological and social adaptations (Ramos, 2006).

That said, displacements of people can be voluntary or forced (Jackson, 1991). The names themselves already portray what each type of migration consists of and therefore no long explanations are needed. In this way, in a simple way, since it is a voluntary migration, the individual leaves his/her country of his/her own free will. On the other hand, if such abandonment is due to factors such as the existence of conflicts or wars, the increasing extremity of weather conditions, the lack of freedom of expression or unemployment and low wages, individuals may feel pressured to leave their country of origin and look for better living and working conditions elsewhere, in other words, this is a forced migration.

Migratory movements, whether internal or international, have an increasingly significant impact on the dynamics of any country or society, not only due to their large-scale growth, as mentioned above, but also due to their ability to be driving forces of social transformation anywhere in the world. Therefore, they help very positively in the development of both countries of origin and destination (Ferreira, 2017). However, this issue and its contours generate a lot of controversy, reason why, before explaining why, it is worth paying attention to the statement by William Lacy Swing (2017, referred by Ferreira, 2017, p. 5), "It is necessary to balance national security with the security of the people (…). Migration policies are generally outdated and look at migration in a static way and not in terms of human mobility. In a globalizing world, we have the free movement of money, services and goods, but we don't have the free movement of people who make it all happen."

Through the above statement, it is clear that the General Director of the International Organization for Migration considers that migration is still a very limited and restricted process. That is, that it does not flow freely. And, this is one of the reasons that gives rise to discussion around this subject. The point is this: if globalization stimulates migrations and these cause enormous positive repercussions in countries, why do they, which on top of that attract migrants, place barriers to their entry? According to Martine (2005, p. 8) "Human capital is a production factor, which, formally, does not have free transit across borders these days; there is no 'global labor market'. Borders open up to the flow of capital and goods, but they are increasingly closed to migrants: this is the big inconsistency (…) with regard to international migrations."

Thus, in this sense, so that all the beneficial potential of migratory processes is explored and exploited and, in an attempt to mitigate the incongruity noted above, it is necessary that countries adopt a set of coherent and favorable policies for the integration of migrants and development.

2.2 Immigration

Immigration is one of the types of migration and is the third major force in population change (Bongaarts & Bulatao, 2000). The International Organization for Migration (2009) defines immigration as the movement of people to a country other than their own with the intention of settling there and, according to the European Commission (undated), this movement should last for a period not less than twelve months. In this sense, immigration can be temporary or permanent. The temporary is the one that is limited in time and the permanent one that occurs when the immigrant is legally allowed to settle in the host country (Organização International para as Migrações, 2009).

Despite the causes that motivate the act of migrating, the immigrant is traditionally seen as "a problem to be solved" (Rodrigues et al., 2013, p. 98), since, in most cases, these people that enter in a country other than their own, aiming for a better or more stable life, have been associated with negative situations, such as crime, marginality or lack of capacity and financial conditions.

The above is in line with what research has shown regarding the way immigrants are positioned in the labor market, which corresponds to an unfavorable and economically less privileged position. It is more frequent to observe unemployed immigrants or, when employed, it is verified that they have lower wages and a lower salary evolution compared to individuals from the country in question with the same age and with the same degree of educational qualifications and professional experience (Ward & Masgoret, 2007) or even inferior. This is because it is difficult for companies to recognize the increasingly developed skills of the immigrant workforce, which leads many to find themselves forced to accept jobs where their skills are not valued. That is, there is an underutilization of them (Viana, 2016).

2.3   Immigration and Labor Market

The study of the relationship between immigrants and the labor market is fundamental for understanding what was mentioned at the end of the previous point. For this, one of the most relevant theories that explains the phenomenon of migrations is presented.

2.3.1   Segmented Labor Market or Dual Labor Market Theory

The theory of the segmented labor market or the dual labor market is one of the best known theoretical models developed on the immigration-labor market relationship. This theory categorizes the labor market into two main segments, the primary and the secondary, and establishes that it is segmented according to different functions, for which candidates are chosen according to specific characteristics, of which the economic, social, among others, with different salary levels (Fiuza-Moura et al., 2014).

In this way, the primary market is characterized by having stable working conditions, good wages, prospects for career advancement within the company (either by transferring to other functions or by expanding skills), guarantees of social protection and also a high social status (Peixoto, 2008). In turn, regarding the secondary market, it can be said that it is the opposite, that is, it presents itself as a market where job insecurity prevails and, consequently, a high turnover rate, low wages (compared to the primary labor market), weak or virtually no opportunities for career advancement, both in terms of promotions and in terms of training, the lack of social protection and, finally, a very low social status (Peixoto, 2008).

This trend toward segmentation of the labor market intensified from the 1970s onwards, with the emergence of new forms of regulation. These allowed the existence of precarious segments and the need to look for a more flexible labor force to gain even more visibility (Kovács, 2005; Kovács & Castillo, 1998). They also resulted in new forms of work, such as temporary, home-based or part-time work (Peixoto, 2008), jobs marked by uncertainty, insecurity and instability (Scheel et al., 2013).

According to most theoretical models, many of the immigrants are present in the secondary market, that is, they work in jobs that require low qualifications and are poorly paid. Thus, and according to Peixoto (2008, p. 22), "It is the fact that there are activities that exist in a labor market of this type that alienates the majority of national citizens and attracts migrants from poor regions. Natives reject the poor economic rewards and low social status associated with some jobs – those that came to be known as the three D's ('dirty, dangerous and difficult'). Immigrants, for their part, even in poor economic conditions, will be able to improve their previous standard of living or, at least, create expectations of future mobility."

In other words, it is possible to continue to perpetuate a labor market with these less favorable and attractive characteristics, since there are still people who agree and submit to them, in an attempt to be able to succeed and be successful (Lopes, 2014).

Finally, it is important to emphasize that immigrants are increasingly qualified people and have increasingly developed skills, however they continue to be excessively present in the secondary market. Even so, it is noteworthy that, despite this fact, the theory of the segmented labor market states that its attraction is not made only to this labor market (Peixoto, 2008).

2.4 Diversity

The growing acceleration of migratory processes also brings with it an increase in diversity in societies and, in the more specific case of this chapter, in the labor market, that is, we are faced with an increasingly diverse workforce.

The concept of diversity is complex, with a multiplicity of definitions for it. The most restricted cover only certain dimensions, such as ethnicity, gender and race (Correia, 2016). Nevertheless, the concept expanded when it began to be realized that the visible dimensions of diversity, some of which were mentioned above, were not the only ones that contributed to the differentiation between collaborators (Kochan et al., 2003).

Thus, and in this follow-up, Thomas (1996) broadly explained diversity by stating that the concept comprises both the differences and the similarities of people. For Kossek and Lobel (1996), diversity refers to a variety of identity attributes displayed by individuals in a given organizational context. Such attributes may include age, physical condition, personality, values and beliefs, marital status, religion, skills, nationality of origin, educational qualifications, among other characteristics. Thus, according to Kapoor (2011), the expansion of the concept may have been due to the fact that it is pertinent to include in the work environment other dimensions of diversity, as an example, educational background, professional experience or personality type.

Still, something that also caused the emergence of broader explanations of the notion of diversity is related to what Barak (1999, p. 51) mentioned about non-visible differences, that is, "(...) should not be omitted from discussions of diversity, as individuals who are different from the organizational mainstream on those invisible characteristics can also experience discrimination and as a result not be able to utilize their full potential at work. A case in point is sexual orientation that can be invisible if a person chooses to keep it confidential, but can trigger prejudice and discrimination if the information gets out."

However, it should be noted that discriminating is not always synonymous with something necessarily negative (Robbins, 2005), as sometimes, in situations of hiring or promotion, it is necessary to differentiate between the skills of the various possible candidates, either for employment or for career advancement. However, if you are talking about discrimination itself, it can cause a bad working environment in the organization.

That said, Dass and Parker (1996) recognize that a culturally diverse workforce can be a very important competitive factor for the organization, allowing it to achieve new competitive advantages, since individuals from different origins, belonging to other cultures and/or who have gone through a wide range of experiences can bring a great deal of baggage to companies, whether of new ideas or new knowledge (D'Netto & Sohal, 1999), giving them greater creativity, for example.

However, for organizations to have the ability to attract and retain employees from different labor markets and, thus, benefit from the advantages that the existence of diversity can offer, namely those mentioned above, Kandola and Fullerton (1994, cited by D'Netto & Sohal, 1999) state that it is crucial that the recruitment and selection processes are based on organizational and work-relevant criteria and that, in turn, the managers who carry out such processes are qualified to assess these same criteria. However, what happens mostly, and specifically in the case of immigrants, is that their skills are not properly perceived, therefore, they experience difficulties in entering the labor market (García-Almeida & Hormiga, 2016) or when already integrated in a company they do not see their skills and abilities recognized. This means that recruiters and/or selectors are not able to qualify immigrants fairly (Carr et al., 2005).

Thus, managing diversity effectively, although very difficult, is extremely important, as it is the human resources strategies that companies apply in this regard that determine whether they will be able to absorb the full potential value of their job seekers or employees. For example, at the level of recruitment and selection, some of the practices that can contribute to improving the management of workforce diversity are the performance of a job analysis process, an anti-discriminatory selection process and focused on the aspects that really are relevant to the job, the inclusion of advertisements in the ethnic language press to attract other types of candidates and the conduct of selection interviews by more than one manager or person in charge (D'Netto & Sohal, 1999).

That said, promoting the acceptance of differences and the development of diverse talents (Carr-Rufino, 1996) is, since then, something critical that organizations must defend in order to be successful in this area of activity.

2.5 Recruitment

> In the knowledge age we currently live in, people are an organization's most important asset, so betting on them remains the best investment a company can make. (Maria da Glória Ribeiro, 2016, cited by Gonçalves, 2018, p. 6)

Over time, the source of competitive advantage for companies changes. However, an aspect that remains timeless for the organization's success is its employees and the way they work (Pfeffer, 1995). Therefore, it is absolutely crucial that companies find the best talent for their open positions, as the quality of their management will strongly depend on the quality of those same people they manage to attract.

In this way, recruitment is a process that consists in attracting qualified people (in the sense that they fulfill the requirements) for a certain function. According to Almeri et al. (2013), these requirements are understood not only in the form of an adequacy/adjustment to the vacancy, but also to the organization's culture. Therefore, the right person for the organization should not necessarily be understood as the one with the highest qualifications, but the one with competencies (knowledge, skills and attitudes) that fit well with the culture of the organization.

The recruitment process is based on the analysis of functions, which consists of both the description and the specification of functions (Pontes, 2010). Regarding the first, it comprises the description of the tasks, that is, the content of the work to be performed. In turn, the specification covers the requirements that the person needs to have to perform a certain task, such as education, experience, skills, responsibilities (Cunha et al., 2010). That said, companies have drawn the profile they want and need their worker to hold, and the recruitment and selection process can be started (Cardoso, 2010). This initial part of preparing the process is extremely important for the process to be successful and, in this way, for the organization to avoid the costs of an inadequate recruitment.

With regard to recruitment methods, these can be divided into internal or external (Ivancevich, 2007). The former is characterized by filling a vacancy using an existing member of the organization, which can mean a promotion, a transfer (horizontal mobility) or the acquisition of new skills, for example, due to the extinction of some work or occupation. Regarding external recruitment, it involves resorting to the job market outside the company to attract someone. It can be done through advertisements in newspapers or on social networks, recruitment agencies and events, among others (Joshi, 2013).

The option for one or the other type of recruitment depends on the organization's resources and business objectives. This means that, by way of example, if a company intends to pursue a mature strategy, what it will certainly need will be a group of internal employees who know very well the particularities of the business (Schuler & Jackson, 1987). In turn, if the company aspires to a growth strategy, where it seeks to find new products or services and/or enter new markets, recruitment may need to be done from abroad (Miles & Snow, 1985; Schuler & Jackson, 1987).

However, with regard to the hiring of immigrant workers, the support of family and friends of employees prevails, since other forms of recruitment are only used when it comes to very specialized and/or higher category job offers (Carvalho, 2004).

2.6 Selection

After the recruitment comes the selection itself and, despite being distinct processes, they are interconnected. According to Chiavenato (2014), in the selection process there is a filtering of candidates, passing only those who have the characteristics that the organization wants. Thus, selection refers to the process of validating and choosing the candidate(s) who meet the qualifications or characteristics that best fit the open position in the company (Cunha et al., 2010; Ivancevich, 2007).

Therefore, from what was exposed above and what has already been mentioned in the topic related to recruitment, it can be inferred that the selection process involves the development of selection criteria, which are based both on the personality characteristics and cultural adjustment of the candidates, as well as in the analysis and specification of functions, as an adjustment between people and functions must be sought in this (Cunha et al., 2010). Having said that, and, after establishing which attributes the candidates need to have to effectively perform their task (Cunha et al., 2010), the company needs to collect this information about them.

Initially, a preliminary reception is carried out, that is, a screening, for example, through the analysis of the CV, which will enable the professional responsible for carrying out the selection process to immediately eliminate those candidates who do not meet the minimum requirements for the open position and which, consequently, do not fit the profile of the vacancy.

Then, there are other selection tools that help the human resources manager to validate and determine if a particular candidate is the most appropriate for the position he is running for, and the one that is most commonly used is the interview (Chiavenato, 2009; Guimarães & Arieira, 2005). This is characterized by being a conversation where the interviewer seeks to gather information about the interviewee, in an attempt to understand their behavior, skills and aspirations (Caetano & Vala, 2002) and also because it is the methodology that exerts the greatest influence on the final decision regarding candidates.

However, the interview can be imprecise, subjective, biased and easily manipulated (Caetano & Vala, 2002; Chiavenato, 2009), and biases may occur (Smith & Robertson, 1993). The most common biases at the time of the interview are the halo (or horn) effect, the similar to me effect (or similarity or "like me" syndrome), the primacy effect and gut feeling bias (intuition), confirmatory bias, fundamental attribution error, stereotypes, the contrast effect, the beauty effect and, finally, the effect of nonverbal behavior (Anderson & Shackleton, 1993, cited by Oliveira, 2015; Bohnert & Ross, 2010; Cunha et al., 2016; Macan & Dipboye, 1990; Wareing & Stockdale, 1987).

For a better understanding of the biases presented above, they are explained in Table 1.

Given the above, that is, the fact that the interview, being subject to inaccuracies and perceptual distortions, may compromise the effectiveness of the selection processes, the best solution may be to use the combination of these with other selection techniques, the exemplary title, with selection tests. These will then function as complements to the interview (Joshi, 2013) and may include knowledge tests, personality tests, psychomotor tests, group dynamics, work samples and assessment centers, among others.

2.7   Influence of Recruitment and Selection Processes on Immigrants' Access to the Labor Market

Having presented the recruitment and selection processes, we now turn to the study of their influence on immigrants' access to the labor market. In this chapter, despite recognizing that these are two processes that are interconnected, each concept was presented separately to understand the essence of each. However, in this section, as there was no isolated study of the influence of each of them in the revised literature, recruitment and selection will be seen as part of a single process.

To set the tone for exposing existing information on this topic, start by noting the following. According to the theory of human capital, the labor market treats any potential worker equally, as this is based on their skills and abilities (Evans & Kelley, 1991). Therefore, it is assumed that, following the perspective of human capital, those responsible for the recruitment and selection processes will not have the power to influence the employability decisions of immigrants (Almeida et al., 2012). But is this really what happens in reality?

As mentioned earlier when presenting the subject of selection, employers or decision-makers related to recruitment and selection may incur biases, for example, in selection interviews (Anderson, 1992). From this, then, it can be seen that their performance can play a determining role in the employment results of immigrants (Almeida et al., 2015), thus not being guided mainly by the theory of human capital. This fact is also demonstrated by other researches, namely by Almeida et al. (2012), who demonstrated that the tolerance, stereotypes and comfort levels of recruiters and selectors can lead them to choose to hire candidates who are closer to the profile/type of people who are already part of their organizations.

Thus, in order to explain the biases that occur in the selection processes of immigrants, two of the known theories are presented, respectively, the similarity-attraction theory and the social identity theory.

Table 1 Biases explanation

Biases	Explanation
Halo effect (or horn effect)	It happens when the interviewer becomes biased when forming an overall opinion of a candidate taking into account only one characteristic, whether positive or negative, that the candidate has, that is, the possession of an (undesirable) trait makes him believe that all other traits the candidate has are also (un)desirable
Similar effect to me (of similarity or "like me" syndrome)	It occurs whenever the interviewer shows preference for a candidate only because he sees that person as having attributes and interests (style and personality, among others) similar to his own
Primacy effect and gut feeling bias (intuition)	The first impression that the interviewer establishes of the candidate will exert a huge weight in his evaluation and final decision, in addition, these first impressions will lead to the candidate's next information/answers being devalued for the hiring decision
Confirmatory bias	It is verified each time that the interviewer looks for information in a candidate to confirm their prejudices in relation to it, disregarding or interpreting in order to confirm the information that does not meet their perceptions/expectations
Fundamental attribution error	It happens when, when evaluating the behavior of the interviewees, the interviewer tends to underestimate the influence of external/contextual factors and overvalue the influence of internal or personal factors, such as the characteristics or dispositions of these individuals
Stereotypes	It is always observed that the interviewer makes a judgment about the interviewee based on the group to which that person belongs. Thus, it takes into account in its categorization central and non-peripheral characteristics, as an example, ethnicity, religion and nationality, among others
Contrast effect	It comes from the tendency of the interviewer to be influenced by the characteristics of candidates previously observed, that is, when evaluating a candidate against another. So, for example, a candidate is more likely to receive a less favorable review if preceded by very good candidates
Beauty effect	Tendency to rank candidates who are perceived as more attractive than others (appearance)

(continued)

Table 1 (continued)

Biases	Explanation
Effect of nonverbal behavior	It always happens that the interviewer's judgment is influenced by the candidate's body language (eye contact, smile, posture), although this may not be a valid predictor of that candidate's future performance. Additionally, tattoos, weight and mannerisms, among others may also be included here

Source Own elaboration

2.7.1 Similarity-Attraction Theory

This theory argues that people will feel attracted to those they perceive to be similar to themselves (Byrne, 1971). That is, if a person perceives similarities to themselves in terms of attitudes, personality traits or other attributes they will feel more attracted by this target of individuals (Montoya & Horton, 2012). According to Montoya et al. (2012), it is also common for people to associate positive feelings with others they perceive to be similar to them, such as the feeling of liking. And, in turn, this feeling of liking will affect how individuals behave toward those who are perceived as their peers (Roebken, 2010).

That said, and given that candidates' first impressions are formed very quickly in the labor market, the similarity-attraction effect can be particularly relevant (Almeida et al., 2015). As stated by Silveira and Hanashiro (2009, p. 121), "In the absence of more information about the attitudes, way of thinking and acting of others, people tend to base their judgments on the manifest characteristics of the other, such as gender, race, appearance, based on symbolism and previous experiences associated with these characteristics, prototypically representing groups of people who have common characteristics, as belonging to a certain social category."

Thus, one of the most popular examples of the similarity-attraction theory is the one proposed by Schneider, the so-called attraction-selection-friction model. According to this model, organizations only attract, select and retain individuals who fit the organization's profile—"(…) on the basis of a similarity to the status quo (…)" (Coates & Carr, 2005, p. 579)—however, this leads companies to homogeneity among their members instead of diversity (Schneider et al., 2001). And thus, immigrants, as they are seen as having certain characteristics that differentiate them from the majority of the organizational sector, end up suffering biases in the selection process (Collier & Burke, 1986, cited by Coates & Carr, 2005).

According to Ben-Ner et al. (2009), ethnicity, religion and clothing influence the way we perceive ourselves and others. That is, they contribute to the accentuation of the similarity effect. However, Almeida et al. (2015) state that the studies carried out in relation to recruitment in this field are very small and limited and, in this sense, they developed a research, with the aim of exploring this gap, which addressed the role played by accent, clothing, non- Anglo-Saxons and by religion in the process of

Table 2 References to other studies developed[1]

Attribute	Results and conclusions of the studies
Accent	• Hosoda and Stone-Romero (2010) found that depending on their accent, applicants for low-status and high-status jobs in the US were treated differently in selection interviews; • Deprez-Sims and Morris (2010) found that when two people with similar qualifications applied to a job, the one with a non-Anglo-Saxon accent was less likely to be perceived as suitable for the job and, as a result, was less likely to be hired » Thus, they demonstrated that foreign accents influence employment decisions
Clothing	• Kang et al. (2011) suggested that those responsible for the recruitment and selection processes prefer candidates who dress in accordance with the image that the organization's current employees present » That said, candidates who present themselves at job interviews in non-Western attire may be treated unfavorably, as they are perceived to be part of an outside group
Non-Anglo-Saxon Names	• Carlsson and Rooth (2007) carried out an experiment, in Sweden, which consisted of sending CVs, for open job vacancies, in which the only difference existing in them was the candidate's name. They found that candidates with names that sounded Arab or Muslim were significantly less likely than other candidates to be called for an interview » Thus, recruiters incurred a form of implicit discrimination and stereotyping toward candidates who had Arabic names, that is, the name of immigrants influences the assessment made by decision-makers in recruitment and selection processes
Religion	• Ghumman et al. (2013) explained that in hiring processes there may be prejudices in relation to religious groups, as certain candidates' religious beliefs can represent challenges for the proper functioning of the organization. This is especially true of observable religious beliefs, such as those in which a religious cap is worn. (Ball & Haque, 2003)

Source Own elaboration

recruitment and selection of immigrants. Thus, they came to the conclusion that all these characteristics/attributes negatively influenced the recruitment processes and, consequently, the decision to hire these individuals. Finally, they also argued that the similarity effect was clearly proven through this research.

In Table 2, there are examples of other studies on these attributes, which had already been carried out by other researchers, and which obtained the same results as the research presented above.

---

[1] All of the researchers' studies presented in Table 2 are cited in Almeida et al. (2015).

2.7.2   Social Identity Theory

Social identity theory is directly related to similarity-attraction theory, as it is often used to describe it. According to this theory, individuals tend to join a group (Tajfel & Turner, 1979) and this "act" of categorizing individuals into groups alone is enough to trigger discrimination between their various members (Tajfel et al., 1971).

However, as reported by Ben-Ner et al. (2009), it is important to note that this trend is characterized as an innate drive for most people. According to Silveira et Hanashiro (2009), people based on the categorizations they make of themselves and others and on the identification or not with such traits will determine their degree of belonging to a particular social group. Thus, in this way, individuals are more likely to approach those who perceive that they share certain characteristics with them, that is, elements belonging to the same group.

However, if the degree of similarity perceived is too high, this can be very harmful for people who do not fit into this group, as they can even be subjected to hatred (Almeida et al., 2015). Indeed, this reality was verified in the immigrant selection processes (Coates & Carr, 2005). One of the possible explanations for this is the fact that immigrant job seekers are seen as different, as mentioned above, and that these differences can act as a threat to groups already formed. In other words, and translating the above into the organizational context, the person responsible for taking decisions on recruitment and selection may see the characteristics of immigrants as something disturbing the degree of similarity that exists between the various members of the organization.

3   Conclusion

In this chapter, it was clear that although immigrants can represent an economic advantage for countries and a competitive advantage for companies, as they allow them to have a more diverse workforce and, consequently, new ideas, knowledge and visions can emerge, being the organizational culture also enriched, they are often viewed negatively by residents of the countries to which they move.

This is due to a belief that people have that immigrants will take their jobs or make wages go down, and also due to the idea of criminality, marginality or lack of skills associated with the same. Thus, this final point is, since then, reflected in the fact that a theory is presented that assumes the existence of two segments of the labor market, the primary and the secondary, with immigrants being mostly present in the latter, that is, where people with low qualifications and poorly paid work, regardless of the reality that they are people with more and more skills. Thus, it is clear that this is immediately an aspect that conditions immigrants' access to a job where their abilities are recognized and valued.

Therefore, it is extremely essential that the recruitment and selection processes are oriented toward the aspects that are really important for the position that the individual is applying for. Thus, this implies that those responsible for these processes are also qualified to assess only those criteria that are considered relevant, that is, they must be able, mainly, to perceive in a fair and adequate way the competences of immigrants. However, what happens most of the time is that this does not happen, since recruiters and selectors are influenced by the perceptual distortions that can occur, for example, at the time of the interview, as an illustration, the final hiring decisions can be affected by the preconceived ideas they have about a particular social group.

That said, this chapter presents two of the theories that prove what was exposed in the previous paragraph, the theories of similarity-attraction and social identity. Through the first and according to the attraction-selection-attrition model, it was understood that in an organizational context only candidates who resemble/fit the organization's profile will be attracted, selected and retained. In this way, immigrants, because they are perceived as different, will certainly be harmed in these processes. Additionally, other aspects that accentuate this perception and, therefore, negatively influence the recruitment and selection processes, refer to issues of accent, clothing, non-Anglo-Saxon names and religion. Regarding the second theory discussed, this is directly linked to the first and explains that individuals tend to join a group, that is, to get closer to people who share certain traits with them. Therefore, once again, those responsible for hiring decisions may see the characteristics of immigrants, which for them are different from those of other organizational members, as explained in this work, as a threat to the similarity, uniqueness that already exists in companies and, as a consequence, the final decision will not be in their favor.

In short, after carrying out this literature review on the influence of recruitment and selection processes on immigrants' access to the labor market, it is concluded that, in fact, these processes are not yet fully optimized and that the people who carry them out can play a decisive role in the decision to hire immigrant candidates. Therefore, it is essential that human resources management professionals avoid or discard the attraction-selection-attrition model, in order to develop various cultures in the organization. However, for this it is also extremely essential that they are aware of how their beliefs, attitudes and stereotypes can influence their behavior in the recruitment and selection processes (Shen et al., 2009), for example, through training or awareness-raising actions for such aspects.

References

Almeida, S., Fernando, M., Hannif, Z., & Dharmage, S. C. (2015). Fitting the mould: The role of employer perceptions in immigrant recruitment decision-making. *The International Journal of Human Resource Management, 26*(22), 2811–2832. https://doi.org/10.1080/09585192.2014.100 3087

Almeida, S., Fernando, M., & Sheridan, A. (2012). Revealing the screening: Organisational factors influencing the recruitment of immigrant professionals. *The International Journal of Human Resource Management, 23*(9), 1950–1965. https://doi.org/10.1080/09585192.2011.616527

Almeri, T. M., Martins, K. R., & de Paula, D. S. P. (2013). O uso das redes sociais virtuais nos processos de recrutamento e seleção. *Revista ECCOM, 4*(8), 77–94. http://unifatea.com.br/seer3/index.php/ECCOM/article/view/557/508

Anderson, N. R. (1992). Eight decades of employment interview research: A retrospective meta-review and prospective commentary. *European Work and Organizational Psychologist, 2*(1), 1–32. https://doi.org/10.1080/09602009208408532

Barak, M. E. M. (1999). Beyond affirmative action: Toward a model of diversity and organizational inclusion. *Administration in Social Work, 23*(3–4), 47–68. https://doi.org/10.1300/J147v23n0 3_04

Ben-Ner, A., McCall, B. P., Stephane, M., & Wang, H. (2009). Identity and in-group/out-group differentiation in work and giving behaviors: Experimental evidence. *Journal of Economic Behavior & Organization, 72*(1), 153–170. https://doi.org/10.1016/j.jebo.2009.05.007

Bohnert, D., & Ross, W. H. (2010). The influence of social networking web sites on the evaluation of job candidates. *Cyberpsychology, Behavior, and Social Networking, 13*(3), 341–347. https://doi.org/10.1089/cyber.2009.0193

Bongaarts, J., & Bulatao, R. A. (2000). *Beyond six billion: Forecasting the world's population.* National Academy Press.

Byrne, D. (1971). *The attraction paradigm.* Academic Press.

Caetano, A., & Vala, J. (2002). *Gestão de recursos humanos. Contextos, processos e técnicas.* Editora RH.

Cardoso, A. (2010). *Recrutamento e seleção de pessoal.* Edições Lidel.

Carr-Ruffino, N. (1996). *Managing diversity: People skills for a multicultural workplace.* Thomson Executive Press.

Carr, S. C., Inkson, K., & Thorn, K. (2005). From global careers to talent flow: Reinterpreting 'brain drain.' *Journal of World Business, 40*(4), 386–398. https://doi.org/10.1016/j.jwb.2005.08.006

Carvalho, L. X. (2004). *Impacto e reflexos do trabalho imigrante nas empresas portuguesas: Uma visão qualitativa.* Alto-Comissariado para a Imigração e Minorias Étnicas.

Chiavenato, I. (2009). *Desempenho humano nas empresas: Como desenhar cargos e avaliar o desempenho para alcançar resultados.* Manole.

Chiavenato, I. (2014). *Gestão de pessoas: O novo papel dos recursos humanos nas organizações.* Manole.

Coates, K., & Carr, S. C. (2005). Skilled immigrants and selection bias: A theory-based field study from New Zealand. *International Journal of Intercultural Relations, 29*(5), 577–599. https://doi.org/10.1016/j.ijintrel.2005.05.001

Comissão Europeia. (s.d.). *Glossário.* Accessed in 16 November 2020 from: https://ec.europa.eu/immigration/glossario_pt-pt

Connell, J., & Burgess, J. (2009). Migrant workers, migrant work, public policy and human resource management. *International Journal of Manpower, 30*(5), 412–421. https://doi.org/10.1108/014 37720910977625

Correia, S. A. I. N. (2016). *A diversidade cultural como uma vantagem para a organização* [Tese de Mestrado, Universidade de Lisboa]. Repositório da Universidade de Lisboa. http://hdl.handle.net/10400.5/13209

Cunha, M. P., Rego, A., Cunha, R. C., Cardoso-Cabral, C., Marques, C. A., & Gomes, J. F. S. (2010). *Manual de gestão de pessoas e do capital humano.* Edições Sílabo.

Cunha, M. P., Rego, A., Cunha, R. C., Cardoso-Cabral, C., & Neves, P. (2016). *Manual de comportamento organizacional e gestão.* Editora RH.

Dass, P., & Parker, B. (1996). Diversity: A strategic issue. In E. E. Kossek & S. A. Lobel (Eds.), *Managing diversity: Human resource strategies for transforming the workplace* (pp. 365–391). Blackwell.

Diário da República Eletrónico. (s.d.). *Declaração Universal dos Direitos Humanos.* Accessed in 8 November 2020 from: https://dre.pt/declaracao-universal-dos-direitos-humanos

D'Netto, B., & Sohal, A. S. (1999). Human resource practices and workforce diversity: An empirical assessment. *International Journal of Manpower, 20*(8), 530–547. https://doi.org/10.1108/014377 29910302723

El-Cherkeh, T. (2009). International labour migration in the EU: Likely social and economic implications. In H. S. Geyer (Ed.), *International handbook of urban policy, volume 2: Issues in the developed world* (pp. 245–258). Edward Elgar.

Evans, M., & Kelley, J. (1991). Prejudice, discrimination, and the labor market: Attainments of immigrants in Australia. *American Journal of Sociology, 97*(3), 721–759. https://www.jstor.org/stable/2781782

Ferreira, P. (2017). *Migrações e Desenvolvimento.* Fundação Fé e Cooperação. https://www.fec ongd.org/pdf/publicacoes/estudoMigracoes_coerencia.pdf

Fiuza-Moura, F. K., Nakatani-Macedo, C. D., de Souza, S. D. C. I., & Maia, K. (2014). *Capital humano e segmentação no mercado de trabalho: Uma análise da indústria paranaense, por níveis de intensidade tecnológica.* https://www.researchgate.net/publication/281069745_CAP ITAL_HUMANO_E_SEGMENTACAO_NO_MERCADO_DE_TRABALHO_UMA_ANA LISE_DA_INDUSTRIA_PARANAENSE_POR_NIVEIS_DE_INTENSIDADE_TECNOL OGICA

García-Almeida, D. J., & Hormiga, E. (2016). Managers' perceptions of the impact of the immigrant workforce: The case of the hotel industry on Lanzarote. *Journal of Human Resources in Hospitality & Tourism, 15*(4), 365–387. https://doi.org/10.1080/15332845.2016.1148565

Gonçalves, C. B. R. (2018). *Implementação de um processo integrado de recrutamento, seleção e acolhimento na Hubel* [Tese de Mestrado, Universidade do Algarve]. Sapientia. http://hdl.han dle.net/10400.1/12205

Guimarães, M. F., & Arieira, J. D. O. (2005). O processo de recrutamento e seleção como uma ferramenta de gestão. *Revista de Ciências Empresariais da UNIPAR, 6*(2), 203–214. https://rev istas.unipar.br/index.php/empresarial/article/viewFile/309/280

Ivancevich, J. M. (2007). *Gestão de recursos humanos.* McGrawHill.

Jackson, J. A. (1991). *Migrações.* Escher.

Joshi, M. (2013). *Human resource management.* Bookboon. https://bookboon.com/pt/human-res ource-management-ebook

Kapoor, C. (2011). Defining diversity: The evolution of diversity. *Worldwide Hospitality and Tourism Themes, 3*(4), 284–293. https://doi.org/10.1108/17554211111162408

Kochan, T., Bezrukova, K., Ely, R., Jackson, S., Joshi, A., Jehn, K., Leonard, J., Levine, D., & Thomas, D. (2003). The effects of diversity on business performance: Report of the diversity research network. *Human Resource Management, 42*(1), 3–21. https://doi.org/10.1002/hrm. 10061

Kossek, E. E., & Lobel, S. A. (1996). *Managing diversity: Human resource strategies for transforming the workplace.* Blackwell.

Kovács, I. (2005). *Flexibilidade de emprego: Riscos e oportunidades.* Celta Editora.

Kovács, I., & Castillo, J. J. (1998). *Novos modelos de produção: Trabalho e pessoas.* Celta Editora.

Lopes, T. P. A. (2014). *Migrações: Novas realidades. Viver num mundo em movimento* [Tese de Mestrado, Universidade de Lisboa]. Repositório da Universidade de Lisboa. http://hdl.handle. net/10451/16139

Macan, T. H., & Dipboye, R. L. (1990). The relationship of interviewers' preinterview impressions to selection and recruitment outcomes. *Personnel Psychology, 43*(4), 745–768.

Martine, G. (2005). A globalização inacabada: Migrações internacionais e pobreza no século 21. *São Paulo em Perspectiva, 19*(3), 3–22. https://doi.org/10.1590/S0102-88392005000300001

Massey, D. S., Arango, J., Hugo, G., Kouaouci, A., Pellegrino, A., & Taylor, J. E. (1999). *Worlds in motion: Understanding international migration at the end of the millennium.* Oxford Clarendon Press.

Milanovic, B. (1999). On the threshold of the third globalization: Why liberal capitalism might fail? *SSRN Electronic Journal*, 1–18. https://doi.org/10.2139/ssrn.262176

Miles, R. E., & Snow, C. C. (1985). Designing strategic human resources systems. *Organizational Dynamics, 13*(1), 36–52. https://doi.org/10.1016/0090-2616(84)90030-5

Montoya, R. M., & Horton, R. S. (2012). A meta-analytic investigation of the processes under-lying the similarity-attraction effect. *Journal of Social and Personal Relationships, 30*(1), 64–94. https://doi.org/10.1177/0265407512452989

Nações Unidas. (2019, novembro 27). *Número de migrantes internacionais no mundo chega a 272 milhões.* Accessed in 8 November 2020 from: https://news.un.org/pt/story/2019/11/1696031

Oliveira, T. C. (2015). *Rethinking interviewing and personnel selection.* Palgrave Macmillan. https://www.palgrave.com/gp/book/9781137497338

Organização Internacional para as Migrações. (2009). Glossário sobre migração. *Direito Inter-nacional da Migração, 22,* 1–87. https://www.acm.gov.pt/documents/10181/65144/Gloss%C3%A1rio.pdf/b66532b2-8eb6-497d-b24d-6a92dadfee7b

Peixoto, J. (2008). Imigração e mercado de trabalho. *Migrações, 2,* 1–215. https://www.om.acm.gov.pt/documents/58428/183863/migracoes2_completo.pdf/e9f06a42-7432-4b98-b310-74112d514491

Pfeffer, J. (1995). Producing sustainable competitive advantage through the effective management of people. *Academy of Management Perspectives, 9*(1), 55–69. https://doi.org/10.5465/ame.1995.9503133495

Pontes, B. (2010). *Planejamento, recrutamento e seleção de pessoal.* Edições LTr.

Ramos, N. (2006). Migração, aculturação, stresse e saúde: Perspectivas de investigação e de intervenção. *Psychologica, 41,* 329–350. http://hdl.handle.net/10400.2/6833

Robbins, S. P. (2005). *Comportamento organizacional.* Pearson Prentice Hall.

Rodrigues, D., Correia, T., Pinto, I., Pinto, R., & Cruz, C. (2013). Um Portugal de imigrantes: Exercício de reflexão sobre a diversidade cultural e as políticas de integração. *Da Investigação às práticas, 4*(1), 86–109. http://hdl.handle.net/10400.21/3424

Roebken, H. (2010). Similarity attracts: An analysis of recruitment decisions in academia. *Educa-tional Management Administration & Leadership, 38*(4), 472–486. https://doi.org/10.1177/1741143210368264

Scheel, T. E., Rigotti, T., & Mohr, G. (2013). HR practices and their impact on the psychological contracts of temporary and permanent workers. *The International Journal of Human Resource Management, 24*(2), 285–307. https://doi.org/10.1080/09585192.2012.677462

Schneider, B., Smith, D. B., & Paul, M. C. (2001). P-E fit and the attraction-selection-attrition model of organizational functioning: Introduction and overview. In M. Erez, U. Kleinbeck, & H. Thierry (Eds.), *Work motivation in the context of a globalizing economy* (pp. 231–246). Lawrence Erlbaum Associates. https://books.google.pt/books?id=kd7IYC4-o3sC&pg=PT292&lpg=PT292&dq=P%E2%80%93E+fit+and+the+attraction-selection-attrition+model+of+organizational+functioning:+Introduction+and+overview.&source=bl&ots=CA-mkD956R&sig=ACfU3U1JTXPEH-sCA2MhorTomXSJ-xUkDQ&hl=pt-PT&sa=X&ved=2ahUKEwjr_tnh4IDyAhWITxUIHbtcC1IQ6AEwBnoECAoQAw#v=onepage&q=P%E2%80%93E%20fit%20&f=false

Schuler, R. S., & Jackson, S. E. (1987). Linking competitive strategies with human resource manage-ment practices. *Academy of Management Executive, 1*(3), 207–219. https://doi.org/10.5465/ame.1987.4275740

Shen, J., Chanda, A., D'Netto, B., & Monga, M. (2009). Managing diversity through human resource management: An international perspective and conceptual framework. *The International Journal of Human Resource Management, 20*(2), 235–251. https://doi.org/10.1080/09585190802670516

Silveira, N. S. P., & Hanashiro, D. M. M. (2009). Similaridade e dissimilaridade entre superiores e subordinados e suas as implicações para a qualidade da relação diádica. *Revista de Administração Contemporânea, 13*(1), 117–135. https://doi.org/10.1590/S1415-65552009000100008

Smith, M., & Robertson, I. T. (1993). *The theory and practice of systematic personnel selection.* Palgrave Macmillan. https://www.palgrave.com/gp/book/9780333586525

Tajfel, H., Billig, M. G., Bundy, R. P., & Flament, C. (1971). Social categorization and intergroup behaviour. *European Journal of Social Psychology, 1*(2), 149–177. https://doi.org/10.1002/ejsp.2420010202

Tajfel, H., & Turner, J. C. (1979). An integrative theory of intergroup conflict. In M. J. Hatch & M. Schultz (Eds.), *Organizational identity: A reader* (pp. 56–65). Oxford University Press.

Thomas, Jr. R. R. (1996). *Redefining diversity.* AMACOM.

Viana, A. M. M. (2016). *Integração dos imigrantes de Leste nas organizações - Uma abordagem às políticas e práticas de recursos humanos desenvolvidas* [Tese de Mestrado, Universidade do Minho]. RepositóriUM. http://hdl.handle.net/1822/44697

Ward, C., & Masgoret, A. M. (2007). Immigrant entry into the workforce: A research note from New Zealand. *International Journal of Intercultural Relations, 31*(4), 525–530. https://doi.org/10.1016/j.ijintrel.2007.03.001

Wareing, R., & Stockdale, J. (1987). Decision making in the promotion interview: An empirical study. *Personnel Review, 16*(4), 26–32.

Can Facebook Data Predict the Level of Sustainable Development in EU-27?

Marius Constantin⦿, Jean-Vasile Andrei⦿, Drago Cvijanovic⦿, and Teodor Sedlarski⦿

Abstract The link between sustainability and digital transformation has been a recurrent topic in the literature and has reached new interest peaks at the beginning of the 2020s. In this context, intelligent data management is not optional, but a requirement if the European Union aims to successfully implement the 2030 Agenda for Sustainable Development in a timely manner. Since data is such a valuable asset and tool for assessing, reporting, and predicting success, Facebook is one of the big data providers that can contribute to a better understanding of the interest and openness of the population toward the concepts of sustainability and sustainable development. Since Facebook collects personal data related to preferences and actions specific to sustainable behaviors, an emergent research question is whether Facebook data can predict the sustainable development level in the case of regions and countries. The objective of this research was to explore the interest of EU-27 citizens for the topic of sustainability by resorting to Facebook Audience Insight database and connect the results with the statistics of sustainable development of the EU-27 states. In this book chapter, multiple digital profiles of EU-27 citizens were developed and explained for each country, based on many factors: nationality, age, gender, the intensity, and nature of engagement on the social network platform.

M. Constantin (✉)
Bucharest University of Economic Studies, Bucharest, Romania
e-mail: marius.constantin@eam.ase.ro

J.-V. Andrei
Petroleum-Gas University of Ploiesti, Ploiesti, Romania
e-mail: ajvasile@upg-ploiesti.ro

National Institute for Economic Research 'Costin C. Kiritescu', Romanian Academy, Bucharest, Romania

D. Cvijanovic
University of Kragujevac, Kragujevac, Serbia
e-mail: drago.cvijanovic@kg.ac.rs

T. Sedlarski
St Kliment Ohridski University of Sofia, Sofia, Bulgaria
e-mail: sedlarski@feb.uni-sofia.bg

© The Author(s), under exclusive license to Springer Nature Switzerland AG 2022
C. Machado and J. P. Davim (eds.), *Sustainability and Intelligent Management*,
Management and Industrial Engineering, https://doi.org/10.1007/978-3-030-98036-8_4

Keywords Data as assets · Intelligent data management · Facebook data
modeling · Sustainable development prediction

1 Introduction

The future is heavily grounded on the moral and socio-economic imperative characterized by the need to achieve sustainability on multiple levels, which can be mode possible if understanding that the natural capital, society, and economy are deeply intertwined and, consequently, humankind needs to act responsibly (Barr, 2003; Robèrt et al., 1997).

The means for achieving sustainability have gotten new perspectives in the context of recent significant developments in the fields of digital information, platforms, systems, and networks (Bradley, 2007; De Bernardi et al., 2019; Sutherland & Jarrahi, 2018). By implementing intelligent technologies, even artificial intelligence (Nañez Alonso et al., 2021), it is possible to positively influence on the quality of life (Ordieres-Meré et al., 2020).

Digitalization is constantly gaining more and more capital in the scientific literature as one of the vectors that significantly contributes to achieving higher levels of sustainable development (Beier et al., 2018, 2020; Del Río Castro et al., 2021; Popescu et al., 2020). Ever since the 2030 Agenda for Sustainable Development emerged (United Nations, 2015), digitalization has gained a momentum in the bigger plan of bringing the Sustainable Development Goals (SDGs) closer to reality with respect to all the dimensions of sustainable development: economic, social and environmental (Tretter et al., 2019). Although not spared of challenges, the mix between digitalization and sustainability has been perceived as a winning combination in pursuing actions meant to contribute to achieving the SDGs (Gouvea et al., 2018; Constantin et al., 2021).

The information and communications technology (ICT) and what might appear to be its unlimited potential to spur sustainability into socio-economic realities has been recognized in the literature (Añón Higón et al., 2017; Hilty & Aebischer, 2015; Khan et al., 2020; N'dri et al., 2021). Not exempted from minor drawbacks, digitalization might cause, in some particular cases, negative environmental impacts, related to waste–especially electronic waste (World Economic Forum, 2019), unsustainable energy consumption behaviors (Petrariu et al., 2021; Voicu-Dorobanțu et al., 2021) and other types of impact. However, the negative effects ICT could potentially have on the environment pale next to its potential if managed properly and responsibly, in the light of achieving sustainability in various economic sectors (Bifulco et al., 2016; Constantin et al., 2021; Davies & Legg, 2018). As far as digital transformations are concerned, the literature is getting rich on papers arguing that Intelligent Data Management, Big Data, Artificial Intelligence, Blockchain are used in a multidimensional manner: (a) with the purpose of measuring the progress made with respect to SDGs and their corresponding targets (Andrei et al., 2021; Aysan et al., 2021;

Di Vaio et al., 2020; MacFeely, 2019); (b) with the purpose of predicting sustainable development trends and directions, based on human behaviors, attitudes–this is possible by processing digital data, approached as a new asset class (Al-Jarrah et al., 2015; Truby, 2020; Weerakkody et al., 2021).

Intelligent data management has many implications for sustainability, at multiple levels. Early empirical studies dealing with intelligent data management used for sustainable development monitoring were focused on measuring the performance made with regards to SDGs and specific SDG targets (Goodenough et al., 1995, 1998). However, recent studies are much more focused on predicting behavioral changes in relation with the 2030 Agenda for Sustainable Development, by resorting to personal data as an asset, a resource, and predictor of change (How et al., 2020; Nigro et al., 2021; Roblek et al., 2020; Vinuesa et al., 2020).

This study aims to provide answers to the research question that represents the title of this book chapter as well: *Can Facebook Data Predict the Level of Sustainable Development in EU-27?* In order to do so, the interest of EU-27 citizens for the topic of sustainability was analyzed by resorting to the Facebook Audience Insight database from which data were extracted. Results were discussed in relation with the statistics covering the level of sustainable development in the EU-27 states, according to the Sustainable Development Index. Moreover, multiple digital profiles of EU-27 citizens were developed and explained for each state member, based on the intensity and nature of engagement on the social network platform. Lastly, cross-sectional econometric models were constructed with the aim of predicting the level of sustainable development in the EU-27 by resorting to SDG Index and Facebook Audience Insight data.

This book chapter follows a standard structure in the literature. The first section after the *Introduction* was dedicated to elaborating a review of the scientific literature dealing with the topics of data collection, digital platforms, social networks, Facebook and its relevance in the context of intelligent data management and enabler, predictor of sustainable development. The following section in this book chapter covers the methodological ground of the quantitative research carried out with the purpose of answering the research question that is also the title of this book chapter. The results and discussions section includes: (a) an overview of different digital behavioral typologies of EU-27 citizens on Facebook in relation with sustainability-related content; (b) a quantitative analysis of the population structure in the EU-27, based on the preference toward engaging with sustainability-related subjects expressed on Facebook; (c) the construction and robustness analysis of multiple econometric models designed to predict the level of sustainable development in the EU-27 by resorting to SDG Index and Facebook Audience Insight data. In the final section of this book chapter, the main findings of the research were pointed out, as well as the limitations of this study. Finally, further research avenues were proposed.

2   Literature Review

2.1   Data, Data Collection, Digital Platforms, and Social Networks

Recent studies show that data can be considered as one of the world's most valuable resource (Birch et al., 2021; Nolin, 2019; Sivarajah et al., 2017). Constant transformations regarding information and communication technologies push data into a new paradigm that is characterized by the fact that data has an associated economic value (Günther et al., 2017).

Data collection, processing, and tracking are crucial for digitalization. However, the permanent streams of personal data operated through third-party tracking systems contribute to rising number of issues dealing with user privacy, democracy and its role in big data management, consumer digital safety and competition (Ajana, 2020; König et al., 2020). User profiling, designing and creating different behavioral typologies based on collected data is made possible through the use of digital platforms (Sadowski, 2020; Constantin et al., 2021).

The roles of digital platforms have been intensely debated in the literature, many arguing for the benefits of using digital platforms in various economic sectors (Baronian, 2020; Duch-Brown & Rossetti, 2020; Zutshi et al., 2021), while others focus on the negative implications of the use of digital platforms (Nuccio & Guerzoni, 2019; Rhum, 2021; Sun et al., 2021). However, social network online platforms have some particularities that set them apart from other types of digital platforms. In this regard, Table 1 was created with the purpose of reviewing selected studies concerning the different perspectives from the literature on the use and functionalities of social network online platforms: positive or negative.

With all their pros and cons, social network platforms are certainly an imperative in today's society that is only one click or tap away from engaging with friends, businesses, concepts, ideas. Social networks have become a place for advocating for ardent societal issues, such as sustainable development, climate change, food waste, and other similar topics (Gonzalez-Lafaysse & Lapassouse-Madrid, 2016; Constantin et al., 2021; Vraga et al., 2015; Vu et al., 2020; Young et al., 2017).

2.2   Facebook and Intelligent Data Management

According to *Facebook Reports—Fourth Quarter and Full Year 2020 Results*, the social network platform has 2.80 billion monthly active users and 65.71% of those visit Facebook or any app owned by Facebook on a daily basis (Facebook, 2021b). Giving people *"the power to build community and bring the world closer together through technologies designed to connect friends and family, find communities and grow businesses"* is what Facebook claims their mission to be.

Table 1 Empirical studies on the use of social network online platforms

Studies	Findings	Type of effects
Gavino et al. (2018) and Wiese and Akareem (2020)	Companies are heavily dependent on social networks as means of advertising and rapid communication channel with their audience (platform users). Through competition on social network platforms for audience attention, companies and entrepreneurs are pushed to find innovative ways to be more performant	Positive
Liu et al. (2017) and Vadisala and Vatsavayi (2017)	Social networks are sharing user-data with third-party consumers and this can be dangerous because potentially sensitive user information can be used against his/her will. The de-anonymization of the user-data can be made possible and this poses major privacy concerns	Negative
Tawa et al. (2016) and Velásquez et al. (2021, p. 19)	Disinformation, misinformation, racism, and other malicious content are easily spreadable on social network platforms and can escalate quickly beyond the control of the individual. This is possible due to technological gaps in the development of social media platforms or lack of intelligent filtering systems	Negative
Ali et al. (2020), Margherita and Heikkilä (2021), and Melanthiou et al. (2015)	The use of social network sites and platforms (SNSP) covers the recruitment, or rather e-recruitment process, making SNSP a valuable business tool that has got a momentum in HR practices. This digital tool proves to be resilient to times of crisis, such as the one caused by the COVID-19 pandemic. Recruitment is no longer a face-to-face dependent process and SNSP is an enabler in this regard	Positive
Kim and Shen (2020) and Yu et al. (2016, 2018)	Social network platforms come in help for older adults to perceive higher levels of feeling connected to a network of friendship that is easier accessible online, in comparison with other than older non-social network users. Older adults tend to find greater benefits in social networks platforms than younger generations	Positive

(continued)

Table 1 (continued)

Studies	Findings	Type of effects
Pallas et al. (2019), Sabin and Olive (2018), and Sacală et al. (2021)	The use of social network platforms add diversity to teaching methods, which is beneficial, especially for students who prefer to engage in learning activities outside class hours. Alternative online communication channels like Facebook and Slack are more likely to be opted by students who feel more comfortable in the digital space. The use of such channels became an imperative during the COVID-19 pandemic, which has reshaped many educational activities	Positive

Source Authors' development

One of the most valuable assets Facebook has is personal data of its users. Through intelligent management of data (Leung et al., 2018), Facebook personal public user-data act as a good predictor that can help interested parties engage better with their audience through market segmentation (Shao et al., 2015), profiling (Fernandez et al., 2012; van Dam & van de Velden, 2015), targeted advertising (Celebi, 2015; Dehghani & Tumer, 2015). The reasons behind wanting to understand audiences and engage better with them are multiple: commercial and economic, wanting to sell more products, generate more profit (Pöyry et al., 2013; Shih, 2010); political (Bucher, 2012; Calvo et al., 2021; Enli & Skogerbø, 2013); social—especially in the case of NGOs that try to increase the awareness on societal issues (Koïvogui et al., 2020; Waters et al., 2009).

This book chapter complements the existing literature on the topic of Facebook data usage. Two components are situated in the spotlight of the study: (a) the component specific to the actual level of sustainable development in the case of each EU-27 member state; and (b) the component specific to Facebook data—by resorting to Facebook Audience Insight platform and designing multiple digital profiles of EU-27 citizens based on: (i) their predisposition to engage with Facebook sustainability-related content; (ii) age; (iii) gender. This research brings its contribution to the literature with a quantitative study focused on using Facebook personal public user-data in order to explore the relation between the level of sustainable development in the EU-27 and the interest of citizens for the content existing on Facebook on the topic of sustainability—this includes Facebook posts, pages, events, groups, and others.

3 Research Data and Methodology

3.1 Data Sources and Collection

Multiple datasets were used in this research. In order to study the interest of EU-27 citizens for the topic of sustainability, data were taken over from the Facebook Audience Insight platform. According to Facebook (2021), their database provides aggregated information about people connected to Facebook based on multiple filters: age and gender breakdowns, education levels, job titles, relationship statuses, and more. In this research, data was collected in June 2021 according to the following Facebook Audience Insight query: *Interests > Hobbies and activities > Politics and social issues > Sustainability*. This query was run for each EU-27 member state and data was collected per age and gender breakdowns, for the following Facebook-defined age groups: 20–24, 25–29, 30–34, 35–39, 40–44, 45–49, 50–54, 55–59, 60–64, 65 and above. After processing this data, multiple digital profiles of EU-27 citizens were developed, based on the previously explained breakdowns and user engagement on the social network platform. This data was later corroborated with the statistics dealing with the level of sustainable development in the EU-27 states: the Sustainable Development Index. This index is adequate to measure the level of development progress of EU-27 member states in relation with the 2030 Agenda for Sustainable Development (Diaz-Sarachaga et al., 2018). The Sustainable Development Index represents "a global assessment of countries' progress toward achieving the SDGs, a complement to the official SDG indicators and the voluntary national reviews" (Sachs et al., 2020). Data regarding the population of each EU-27 member state were taken over in June 2021 from the official Eurostat database, based on the following online indicator data code: DEMO_PJANGROUP—"Population on 1 January by age group and sex" (Eurostat, 2021).

3.2 Research Methodology

Cross-sectional linear regression models were constructed with the aim of predicting the level of sustainable development in the EU-27 by resorting to the data collected and processed as explained in Sect. 3.1. Consequently, in the designing phase of the econometric models, the SDG Index was always considered the endogenous variable. Multiple exogenous variables were tested, according to the various typologies of identified EU-27 citizens–breakdowns: age, gender, the predisposition to engage on Facebook posts, events, and pages related to sustainability issues. The econometric models were constructed by applying the least squares method in EViews.

4 Results and Discussion

4.1 An Overview of Different Digital Behavioral Typologies on Facebook in Relation with Sustainability-Related Content

Table 2 consists of the main indicators used in the design of digital behavioral typologies of EU-27 citizens according to their engagement on Facebook with sustainability-related content: Facebook posts, pages, events. Indicators were calculated per EU-27 member.

Descriptive statistics for each of the ten indicators analyzed in Table 2 were calculated and included in Table 3. To those ten indicators, the SDG Index was also added in order to provide a better perspective on the relation between the level of sustainable development and different digital behavioral typologies of EU-27 citizens according to their engagement on Facebook with sustainability-related content: Facebook posts, pages, events.

Statistics from Tables 2 and 3 were discussed in the following section at the level of each EU-27 member state, but more in-depth, as the population structure was another factor considered as well.

4.2 Population Structure in the EU-27 Based on the Preference of Citizens Towards Engaging with Sustainability-Related Subjects on Facebook

4.2.1 Austria

Among all of the analyzed age groups, Austrian women between 25 and 29 years old are the most interested in sustainability-related actions, events, subjects, according to Facebook Audience Insight (Fig. 1). If compared to the EU-27 average, referring to the same category (i.e.: 25–29 years old women), Austrian women are more likely to be advocating for sustainability, considering that the percentage of Austrian women with Facebook accounts interested in sustainability-related subjects from the total Austrian women with Facebook accounts represented 5.65%, whereas the EU-27 average was 4.39%. However, in the case of Austrian women, this type of preference was observed only for the three five age groups analyzed, 30–34 years old being the last group included.

In the case of Austrian men, their preference for sustainability is slightly less intense than in the case of women. In this regard, the share of Austrian men with Facebook account, interested in sustainability-related subjects from the total Austrian men that have a Facebook account represented 24.82%, whereas the share of Austrian women with Facebook account, interested in sustainability-related subjects from the

Table 2 List of indicators used in the design of digital behavioral typologies on Facebook

EU-27 Countries	I₁	I₂	I₃	I₄	I₅	I₆	I₇	I₈	I₉	I₁₀
	Population of age 20 and above		Population of age 20 and above that own a Facebook account			Population of age 20 and above with Facebook account, interested in sustainability-related subjects				I₈–I₉ Difference
	Number	Percentage from EU-27 total population	Number	Percentage of population with Facebook account from EU-27 population	Percentage of population that has a Facebook account from the national population	Number	Percentage of population with Facebook account, interested in sustainability-related subjects from EU-27 population	Percentage of population with Facebook account, interested in sustainability-related subjects from the national population that have Facebook accounts	Percentage population with Facebook account that are interested in sustainability-related subjects from the national population	
AT	7.1M	2.01%	2.2M	1.71%	30.64%	1.2M	2.05%	56.14%	17.20%	38.94%
BE	8.9M	2.51%	3.8M	2.96%	42.49%	1.7M	2.82%	44.58%	18.94%	25.64%
BG	5.6M	1.58%	1.9M	1.48%	33.71%	0.8M	1.27%	40.16%	13.54%	26.62%
HR	3.2M	0.92%	1.1M	0.86%	33.56%	0.5M	0.80%	43.57%	14.62%	28.95%
CY	0.6M	0.20%	0.6M	0.44%	81.89%	0.2M	0.42%	44.21%	36.20%	8.01%
CZ	8.5M	2.39%	2.9M	2.26%	34.09%	1.2M	2.06%	42.64%	14.54%	28.11%
DK	4.5M	1.27%	2M	1.56%	44.19%	1.7M	2.82%	84.95%	37.54%	47.41%
EE	1.0M	0.29%	0.4M	0.30%	36.28%	0.2M	0.33%	52.24%	18.95%	33.28%
FI	4.35M	1.22%	1.5M	1.17%	34.42%	1.1M	1.86%	74.67%	25.70%	48.96%
FR	51.1M	14.33%	19M	14.81%	37.18%	6.6M	10.92%	34.55%	12.85%	21.71%
DE	67.8M	19.03%	20M	15.59%	29.48%	9.3M	15.44%	46.44%	13.69%	32.74%
EL	8.6M	2.42%	3.4M	2.65%	39.35%	1.6M	2.75%	48.62%	19.13%	29.49%
HU	7.8M	2.20%	3M	2.34%	38.17%	1.5M	2.48%	49.67%	18.96%	30.71%

(continued)

Table 2 (continued)

EU-27 Countries	I₁	I₂	I₃	I₄	I₅	I₆	I₇	I₈	I₉	I₁₀
	Population of age 20 and above		Population of age 20 and above that own a Facebook account			Population of age 20 and above with Facebook account, interested in sustainability-related subjects				I₈–I₉ Difference
	Number	Percentage from EU-27 total population	Number	Percentage of population with Facebook account from EU-27 population	Percentage of population that has a Facebook account from the national population	Number	Percentage of population with Facebook account, interested in sustainability-related subjects from EU-27 population	Percentage of population with Facebook account, interested in sustainability-related subjects from the national population that have Facebook accounts	Percentage population with Facebook account that are interested in sustainability-related subjects from the national population	
IE	3.6M	1.02%	1.5M	1.17%	41.26%	1.1M	1.83%	73.40%	30.29%	43.11%
IT	49.0M	13.76%	19M	14.81%	38.74%	10.5M	17.39%	55.05%	21.33%	33.72%
LV	1.5M	0.42%	0.5M	0.39%	33.05%	0.3M	0.44%	52.48%	17.34%	35.14%
LT	2.2M	0.63%	0.8M	0.65%	37.49%	0.5M	0.80%	57.05%	21.39%	35.66%
LU	0.4M	0.14%	0.3M	0.17%	44.65%	0.1M	0.14%	39.50%	17.64%	21.86%
MT	0.4M	0.12%	0.3M	0.17%	51.98%	0.1M	0.20%	55.23%	28.71%	26.52%
NL	13.6M	3.82%	5.4M	4.21%	39.61%	2.6M	4.40%	49.04%	19.42%	29.61%
PL	30.3M	8.50%	9.2M	7.17%	30.35%	2.1M	3.42%	22.38%	6.79%	15.59%
PT	8.3M	2.34%	3.6M	2.81%	43.10%	2.1M	3.42%	57.14%	24.62%	32.51%
RO	15.2M	4.28%	5.5M	4.29%	36.01%	2.2M	3.69%	40.36%	14.54%	25.83%
SK	4.3M	1.21%	1.4M	1.09%	32.32%	0.6M	1.02%	43.93%	14.20%	29.73%

(continued)

Table 2 (continued)

EU-27	I_1	I_2	I_3	I_4	I_5	I_6	I_7	I_8	I_9	I_{10}
Countries	Population of age 20 and above		Population of age 20 and above that own a Facebook account			Population of age 20 and above with Facebook account, interested in sustainability-related subjects				I_8–I_9 Difference
	Number	Percentage from EU-27 total population	Number	Percentage of population with Facebook account from EU-27 population	Percentage of population that has a Facebook account from the national population	Number	Percentage of population with Facebook account, interested in sustainability-related subjects from EU-27 population	Percentage of population with Facebook account, interested in sustainability-related subjects from the national population that have Facebook accounts	Percentage population with Facebook account that are interested in sustainability-related subjects from the national population	
SI	1.6M	0.47%	0.6M	0.45%	34.39%	0.2M	0.39%	40.86%	14.05%	26.81%
ES	38.0M	10.67%	15M	11.69%	39.42%	7.3M	12.17%	48.80%	19.24%	29.56%
SE	7.9M	2.22%	3.6M	2.81%	45.43%	2.8M	4.67%	77.97%	35.42%	42.55%

Source Authors' development

Table 3 Descriptive statistics of the indicators

Statistics	SDG Index	I_1	I_2	I_3	I_4	I_5	I_6	I_7	I_8	I_9	I_{10}
Mean	78.591	13.2M	3.70%	4.7M	3.70%	39.38%	2.2M	3.70%	50.95%	20.25%	30.70%
Median	78.110	7.1M	2.01%	2.2M	1.71%	37.49%	1.2M	2.06%	48.80%	18.95%	29.61%
Maximum	84.720	67.8M	19.03%	20M	15.59%	81.89%	10.4M	17.39%	84.95%	37.54%	48.96%
Minimum	74.310	0.4M	0.12%	0.2M	0.17%	29.48%	0.1M	0.14%	22.38%	6.79%	8.01%
Std. Dev	2.996	17.8M	5.01%	6.1M	4.77%	10.00%	2.8M	4.66%	13.78%	7.69%	8.90%
Skewness	0.376	1.831	1.831	1.663	1.663	2.960	1.864	1.864	0.746	0.868	-0.102
Kurtosis	2.465	5.211	5.211	4.310	4.310	13.269	5.319	5.319	3.604	3.083	3.636
J–B*	0.959	20.589	20.589	14.376	14.376	158.044	21.680	21.680	2.913	3.396	0.502
Obs.**	27	27	27	27	27	27	27	27	27	27	27

*J–B = abbreviation for Jarque–Bera; **Obs = Observations; M = abbreviation for million
Source Authors' computation

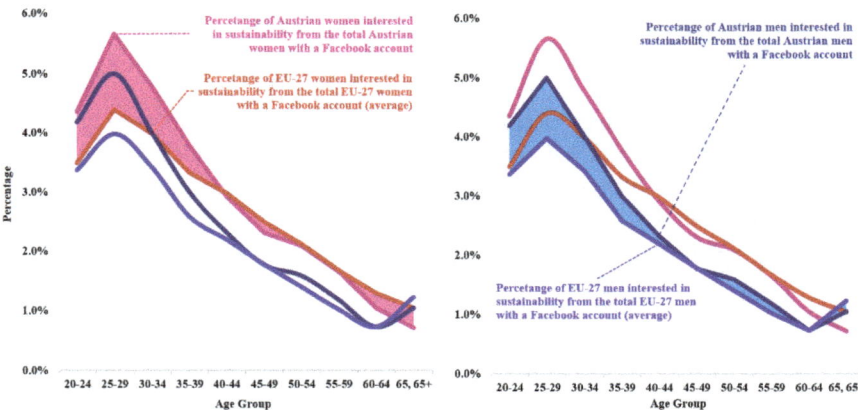

Fig. 1 Austria: Population's interest for sustainability-related subjects. *Source* Authors' development based on Facebook data—Audience Insight (2021)

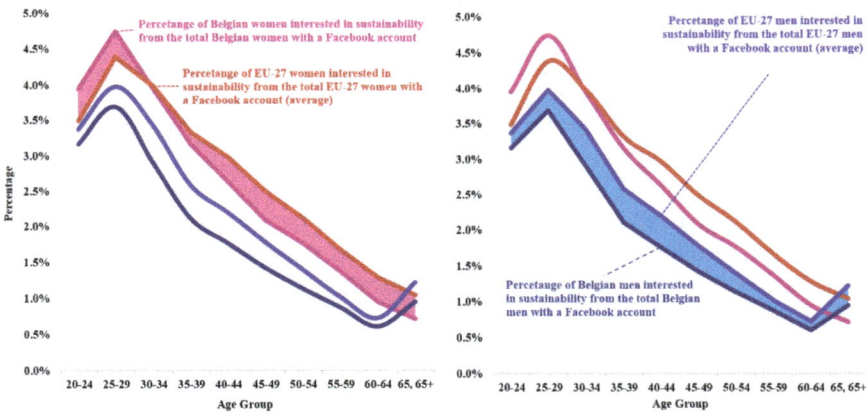

Fig. 2 Belgium: Population's interest for sustainability-related subjects. *Source* Authors' development based on Facebook data—Audience Insight (2021)

total Austrian women that have a Facebook account represented 29.28%. Young Austrian men (20–39 years old) are more likely to express it than another group ages (39–65 + years old).

4.2.2 Belgium

Belgian men are less likely to have Facebook accounts and express their interest in sustainability-related subjects from the total Belgian men that have Facebok accounts (1.86% age group average), compared to EU-27 age group average with the same characteristics (2.17%). Similarly, Belgian women follow the same pattern (Fig. 2), with the difference that Belgian age groups 20–24 (3.95%) and 25–29 (4.74%) are situated above the EU-27 average: 3.49% in the case of 20–24 years old and 4.39% in the case of 25–29 years old.

The greatest discrepancy between Belgian men who are empowering sustainability through the lens of the percentage of Belgian men with Facebook account interested in sustainability-related subjects from all the Belgian men connected to Facebook and the EU-27 average was noticed in the case of the 30–34 years old age group (2.89%—Belgium and 3.41%—EU-27 average).

4.2.3 Bulgaria

Among all of the analyzed typologies of European women, Bulgarian women are those that show little interest in sustainability-related actions, events, subjects, according to Facebook Audience Insight, (Fig. 3). This is supported by the fact that the percentage of Bulgarian women with Facebook account interested in sustainability-related subjects from all the Bulgarian women connected to Facebook was 2.63% in the case of the 25–29 age group, while the EU-27 average for the same age group was 4.39%, therefore a percentage greater with 1.75% that of the Bulgarian women–signaling the lack of interest for the topic of sustainability.

Unlike Bulgarian women, Bulgarian men are more likely to engage with Facebook pages, events, and posts specific to sustainable development. In this regard, it is important to mention that the share of Bulgarian men with Facebook accounts, interested in sustainability-related subjects from the total Bulgarian men that have a Facebook account represented 19.47%, whereas the share of Bulgarian women with Facebook account, interested in sustainability-related subjects from the total Bulgarian women that have a Facebook account represented 17.88%.

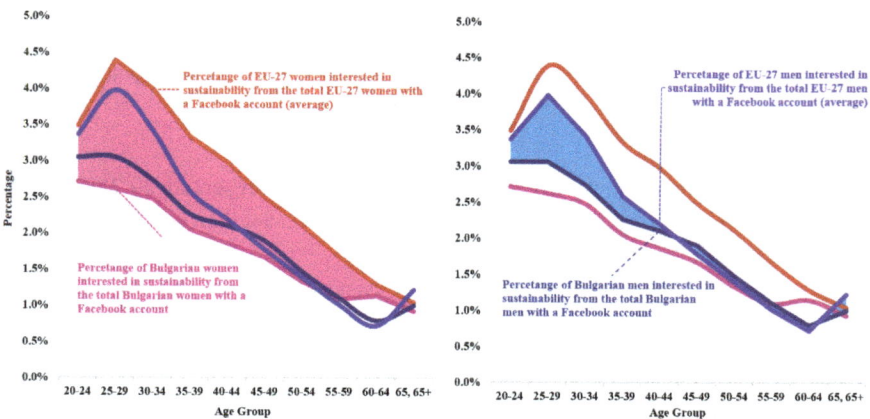

Fig. 3 Bulgaria: Population's interest for sustainability-related subjects. *Source* Authors' development based on Facebook data—Audience Insight (2021)

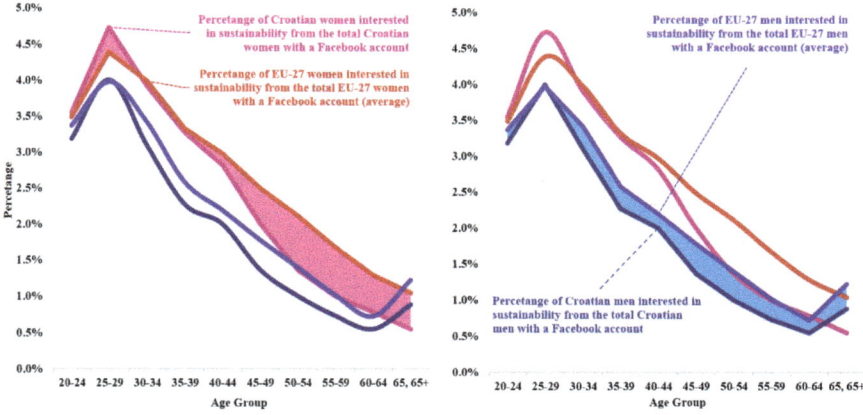

Fig. 4 Croatia: Population's interest for sustainability-related subjects. *Source* Authors' development based on Facebook data—Audience Insight (2021)

4.2.4 Croatia

Croatians' behavior on Facebook with respect to sustainable development pages, events, and posts is almost the exact typology of the average EU-27 citizen (Fig. 4). With a population of 3,278,118 of the age 20 and above (0.92% of EU-27's population), Croatia is among the European countries situated at the bottom of the ranking concerning the percentage population with Facebook accounts with citizens interested in sustainability-related subjects from the total national population (14.62%). Moreover, Croatia (43.57%) is also below the EU-27 (50.95%) average when analyzing the percentage of population with Facebook accounts and interested in sustainability-related subjects from the total national population with Facebook accounts.

Croatian women situated in the 25–29 years old age group represent a segment of the total Croatian population that expressed most their preference for sustainable activities, according to Facebook Audience Insight. In this regard, the share of Croatian women with Facebook accounts that expressed interest in sustainability-related subjects from all the Croatian women connected to Facebook was 4.73% at the moment of carrying out this research (June 2021), while the EU-27 average regarding the same age group and gender was 4.39% (0.34% lower).

4.2.5 Cyprus

Similar with the situation of Croatian people, Cypriots' engagement with Facebook pages, events, and posts demonstrate a concern for sustainable development that is almost the same with the typology identified in the case of the average EU-27 citizen (Fig. 5). From the perspective of the percentage of the population that owns a Facebook account, Cyprus is an EU-27 outlier, since 81.89% of the Cypriot population of

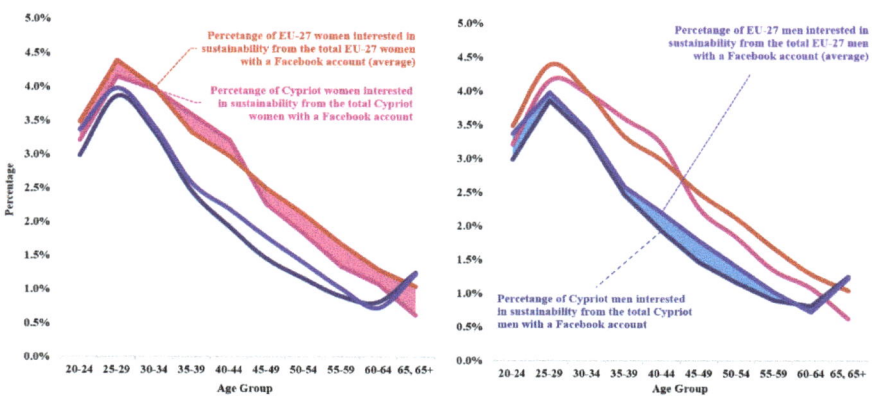

Fig. 5 Cyprus: Population's interest for sustainability-related subjects. *Source* Authors' development based on Facebook data—Audience Insight (2021)

age 20 and above has a Facebook account, while the EU-27 average was 39.38%. Out of the 81.89%, only 44.21% of those (no matter the age and gender) are interested in subjects related to sustainability.

The percentage of Cypriot men with Facebook account who expressed interest in sustainability-related subjects from the total Cypriot men that have a Facebook account represented 20.12%, whereas the share of Cypriot women with Facebook account, interested in sustainability-related subjects from the total Cypriot women that have a Facebook account represented 25.24%, resulting a difference of 5.12% in the favor of Cypriot women, more likely to foster sustainability than men.

4.2.6   Czechia

Young Czech people (20–24 years old) are the ones that manifested the greatest interest on Facebook toward sustainability-related subjects from the total analyzed Czech typologies (Fig. 6). Czechia has a share of 3.7% of EU-27's population aged 20 or above and only 34.09% of Czech people have a Facebook account. Less than half of those (42.64%) manifested any interest for sustainability-related topics on Facebook.

Compared to the EU-27 average men (3.36%) and women (3.49%) citizens aged between 20 and 24 years old, the percentage of Czech women in the same age group with Facebook accounts and who expressed interest in sustainable development topics from the total Czech women that have a Facebook account was 0.85% greater, similar to the case of men (0.43% greater). This kind of preference marked by the concern for sustainability was less observed in the case of Czech citizens older than 35 years old, similar to the case of Austria, Belgium, etc.

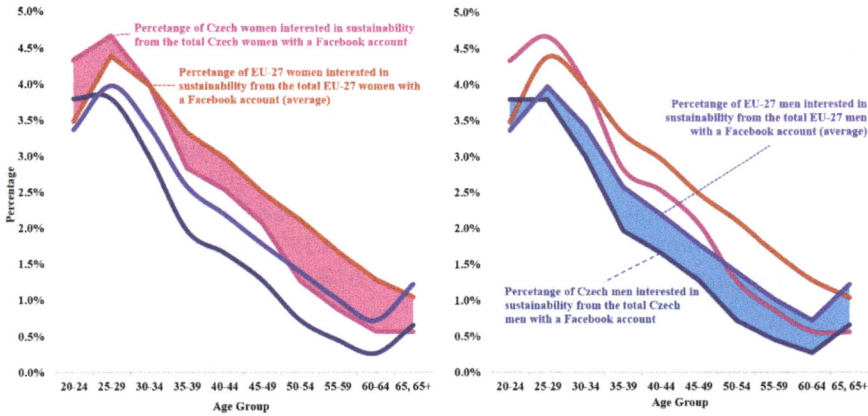

Fig. 6 Czechia: Population's interest for sustainability-related subjects. *Source* Authors' development based on Facebook data—Audience Insight (2021)

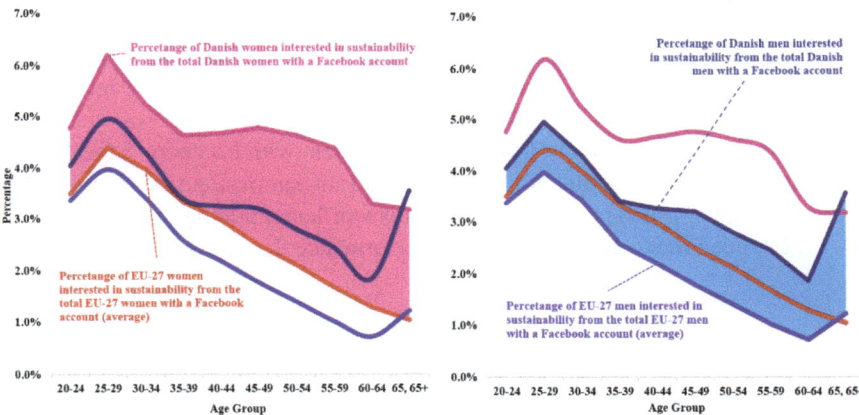

Fig. 7 Denmark: Population's interest for sustainability-related subjects. *Source* Authors' development based on Facebook data—Audience Insight (2021)

4.2.7 Denmark

Denmark is one of the EU-27's leaders when it comes to citizens expressing preferences toward sustainability on Facebook (Fig. 7). The percentage of Danish men with Facebook account who expressed interest in sustainability-related subjects from the total Danish men that have a Facebook account represented 33.80%, whereas the share of Danish women with Facebook account, interested in sustainability-related subjects from the total Danish women that have a Facebook account represented 45.69%, resulting a difference of 12.13% in the favor of Danish men by reporting to EU-27 average and 18.93% if referring to women (EU-27 average). These statistics show that Danish people are more likely to advocate for achieving higher levels of sustainable development more than other European people do on the social network.

Unlike any other typology, Danish women represented by the 55–59 years old age group were a segment of the total Danish population that expressed most their preference for sustainable activities, reported to similar age group results from other EU-27 countries. Regarding this particular age group, the share of Danish women with Facebook accounts that expressed interest in sustainability-related subjects from all the Danish women connected to Facebook was 4.38%, while the EU-27 average regarding the same age group and gender was 1.67% (2.71% lower).

4.2.8   Estonia

Situated below the EU-27 average regarding the percentage of citizens expressing their preferences on Facebook related to sustainability-related topic from the total population (Fig. 8), Estonia is not performing any better when analyzing the percentage of population with Facebook accounts from the total national population (36.28%). However, 52.24% of the Estonians connected to Facebook expressed interest in sustainability-related subjects. If reported to the total population, only 20.25% of the Estonians are connected to Facebook and are interested in sustainability.

Similar with the situation of Cypriots' engagement with Facebook pages, events, and posts dealing with sustainability, Estonians do not demonstrate a significant effort in connecting with such pages and events. It is worth mentioning that the age group 30–34, Estonian men, registered a better performance than the EU-27 average did,

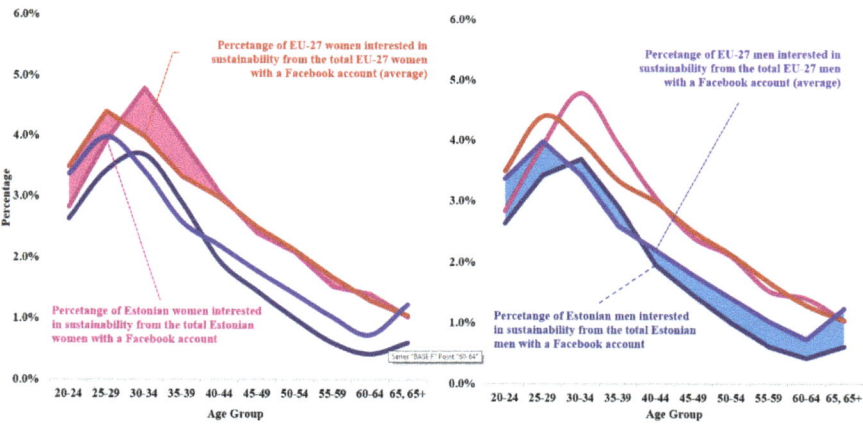

Fig. 8 Estonia: Population's interest for sustainability-related subjects. *Source* Authors' development based on Facebook data—Audience Insight (2021)

as far as the percentage of men with Facebook accounts interested in sustainability-related subjects from all the men connected to Facebook is concerned: 4.78% in the case of Estonian men and 3.98% in the case of EU-28 average (0.80% difference).

4.2.9 Finland

Like Denmark, Finland is another EU-27 leader when analyzing the percentage of citizens expressing preferences toward sustainability on Facebook (Fig. 9). However, with a total population that represents 1.22% out of EU-27's population of age 20 and above, only 34.42% out of the Finnish population have a Facebook account. More than a million Finnish people engaged with Facebook pages, posts, events, etc. and expressed their interest in sustainability, which means that 74.67% out of the Finnish people connected to Facebook are empowering sustainability on the digital platform, representing 1,46 times more than the EU-27 average.

The percentage of Finnish men with Facebook account who expressed interest in sustainability-related subjects from the total Finnish men that have a Facebook account represented 28.67%, whereas the share of Finnish women with Facebook account, interested in sustainability-related subjects from the total Finnish women that have a Facebook account represented 38.63%, resulting a difference of 7.00% in the favor of Finnish men by reporting to EU-27 average and 11.87% if referring to women (EU-27 average). Results confirm that Finnish people strive to connect better with sustainability-related issues than other European citizens do.

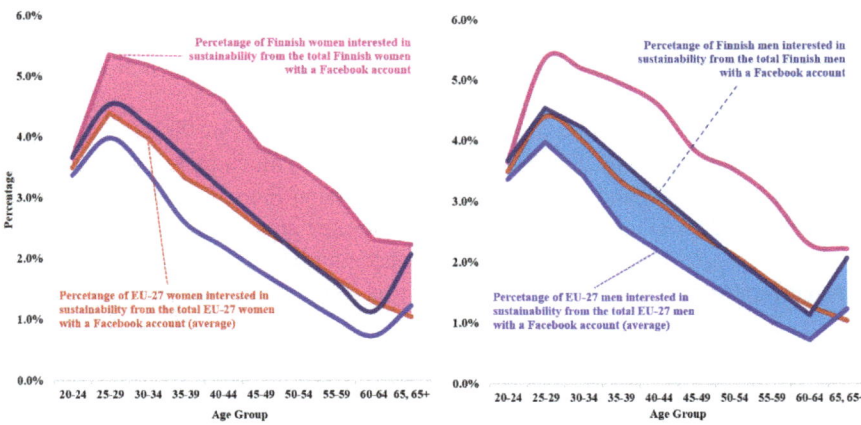

Fig. 9 Finland: Population's interest for sustainability-related subjects. *Source* Authors' development based on Facebook data—Audience Insight (2021)

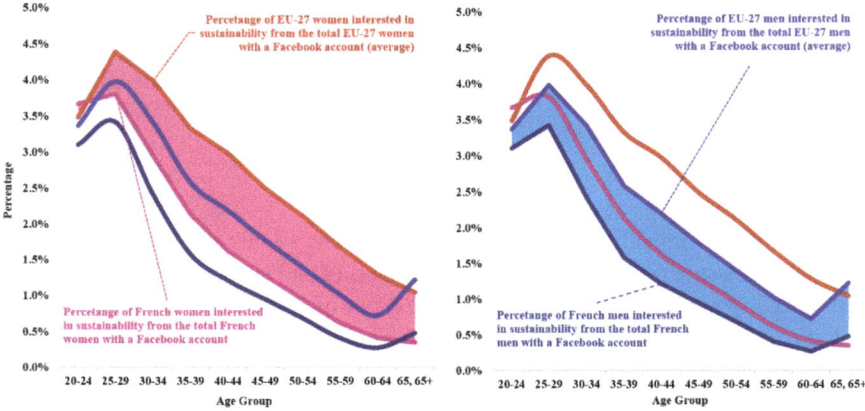

Fig. 10 France: Population's interest for sustainability-related subjects. *Source* Authors' development based on Facebook data—Audience Insight (2021)

4.2.10 France

Unlike Denmark and Finland, France is far from being a EU-27 leader regarding citizens advocating for sustainability on social network platforms–this can be noticed by analyzing the percentage of citizens expressing preferences toward sustainability on Facebook (Fig. 10). Although French population summed up to 14.33% of EU-27's population aged 20 or above, only 37.18% of the French population is connected to Facebook via an account. Unfortunately, the share of French people with Facebook accounts who are interested in sustainability-related subjects from the total French people that own a Facebook account (34.55%) is 16.40% lower than the EU-27 average.

Unlike other European typologies previously analyzed (see Sects. 4.2.1, 4.2.2, 4.2.4) characterized by the fact that younger people are much more prone to engage with sustainability-related content on Facebook, young French people show interest below the EU-27 average, with the exception of the age group 20–24 French men. The 20–24 age group (French men with Facebook accounts) that has associated Facebook metadata confirming the interest for the topic of sustainability gathers 3.67% (above the EU-27 average with 0.18%) of the French men that have Facebook accounts.

4.2.11 Germany

With a slightly better performance than the EU-27 average, Germany shows the signs of an active civil society on Facebook that engages with sustainability-related subjects (Fig. 11). Although it gathers the biggest share of EU-27 population (19.03%), only 29.48% of it owns a Facebook account (the smallest percentage among all the

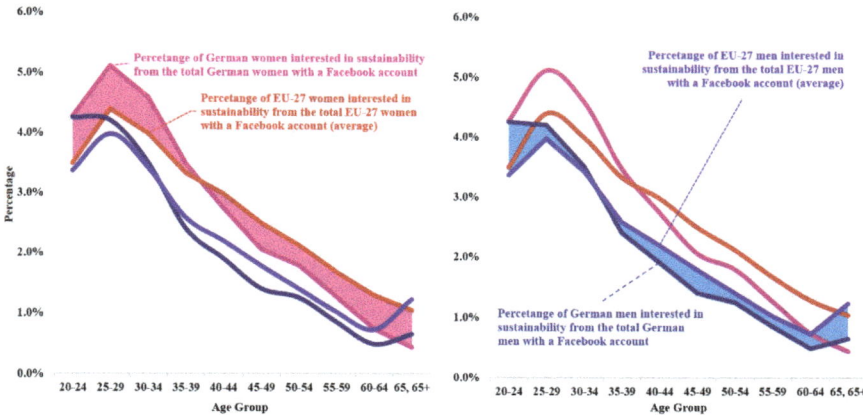

Fig. 11 Germany: Population's interest for sustainability-related subjects. *Source* Authors' development based on Facebook data—Audience Insight (2021)

EU-27 members). However, close to half (46.44%) of German citizens with Facebook accounts engaged with Facebook posts, pages, events dealing with the topic of sustainability, which, once again, is below the average of the EU-27 (50.95%).

A significant behavioral discrepancy was observed between German age groups 20–24, 25–29, 30–34, and the rest of the German age groups. The inflection point based on the age factor shows that German people tend to engage less on Facebook on sustainability-related topics, posts, and pages after the age of 34, especially in the case of German man between 60 and 64 years old. These findings might not implicitly support the fact that older German people are not interested in the topic of sustainability, but rather these findings leave the door open to further research avenues in the direction of the communication channels specific to these categories of citizens.

4.2.12 Greece

Greece shows the signs of a European country with a younger population less interested in sustainability-related issues on Facebook than the 40–54 years old Greeks are (Fig. 12). Analyzing age group averages, Greek citizens with Facebook accounts within the 40–44, 45–49, 50–54 age groups that also have associated Facebook metadata confirming the interest for the topic of sustainability represent, on average, 2.25% (men) and 2.80% (women) of the Greek citizens that have Facebook accounts (while the EU-27 averages for the corresponding age groups are 1.79% in the case of men and 2.53% in the case of women).

Regarding the predisposition to have a Facebook account, almost 40% (39.38%) of the Greek population of age 20 and above are connected to Facebook. Almost half (48.62%) of those connected engaged with Facebook posts, pages, events dealing

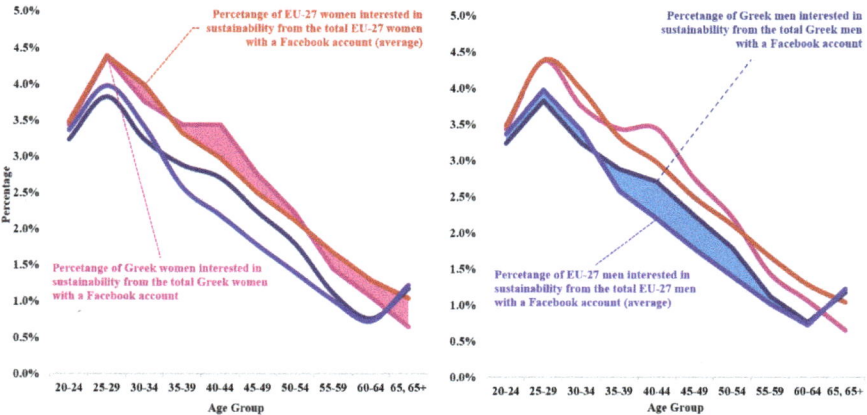

Fig. 12 Greece: Population's interest for sustainability-related subjects. *Source* Authors' development based on Facebook data—Audience Insight (2021)

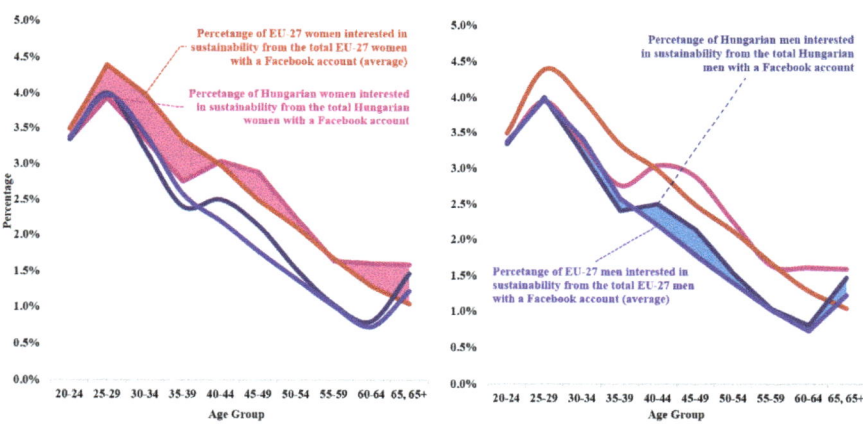

Fig. 13 Hungary: Population's interest for sustainability-related subjects. *Source* Authors' development based on Facebook data—Audience Insight (2021)

with the topic of sustainability, which, is slightly below the EU-27 average (46.44%). If reported to the total Greek population of age 20 and above, connected or not to Facebook, only 20.25% of this population is connected to Facebook and expressed concerns for sustainability on the social platform.

4.2.13 Hungary

Hungarian population follows the same behavioral patterns as explained in Sect. 4.2.12 in the case of Greece. In this regard, Fig. 13 resembles Fig. 12.

Young Hungarians are less interested in sustainability-related issues and engage less on Facebook than the 40–54 years old Hungarians do. Looking at age group averages, Hungarian citizens with Facebook accounts within any age group that includes persons above 39 years old and that have associated Facebook metadata confirming the interest for the topic of sustainability represent, on average, 1.58% (men) and 2.16% (women) of the Hungarian citizens that have Facebook accounts (while the EU-27 averages for the corresponding age groups are 1.39%—men; 1.93%—women).

As far as the likelihood of Hungarians to have a Facebook account is concerned, 38.17% of the Hungarian population of age 20 and above is connected to Facebook. Almost half (49.67%) of the Hungarians connected on the social platforms have also engaged with Facebook posts, pages, events dealing with sustainability-related issues. If reported to the total Hungarian population of age 20 and above, only 18.96% of those are connected to Facebook and expressed their concern for sustainability on the social platform.

4.2.14 Ireland

Ireland's situation is similar to that of Denmark's (Sect. 4.2.7) and Finland's (Sect. 4.2.9).Ireland is also one of the EU-27's leaders when analyzing the willingness of citizens to express their preferences toward sustainability on Facebook (Fig. 14). The percentage of Irish men with Facebook account who expressed interest in sustainability-related subjects from the total Irish men that have a Facebook account

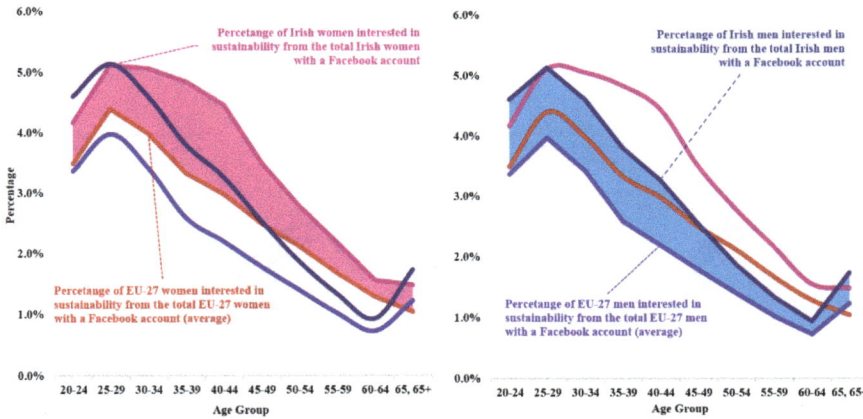

Fig. 14 Ireland: Population's interest for sustainability-related subjects. *Source* Authors' development based on Facebook data—Audience Insight (2021)

represented 29.80%, whereas the share of Irish women with Facebook account, interested in sustainability-related subjects from the total Irish women that have a Facebook account represented 35.10%–therefore resulting a difference consisting of 5.3 percentage points in the favor of Irish women.

Regarding the predisposition of Irish citizens to have a Facebook account, 41.26% of the Irish population of age 20 and above is connected to Facebook. Huge percentages (73.40%) of those connected on the social platform have also engaged with Facebook posts, pages, events touching sustainability-related issues. However, by reporting to the total Irish population of age 20 and above, 30.29% of those are connected to Facebook and have also expressed their concern for sustainability on the social platform.

4.2.15   Italy

Resembling Germany's situation (Sect. 4.2.11), Italy has a slightly better performance than the EU-27 average and results confirm the activity of a Italian vigilant civil society on Facebook, a society that engages with sustainability-related subjects (Fig. 15). Italy gathers one of the biggest share of EU-27 population of age 20 and above (13.71%), out of which 39.48% own a Facebook account. More than half of Italian citizens with Facebook accounts (55.05%) engaged with Facebook posts, pages, events related to the topic of sustainability, making Italy a EU-27 leader in this regard.

Italy shows the signs of a European country with a younger population more interested in sustainability-related issues on Facebook than the older Italian population do. Statistics related to this fact is amplified if referring to the Italian men-women ratio, in the favor of Italian men. Analyzing age group averages, Italian men with Facebook

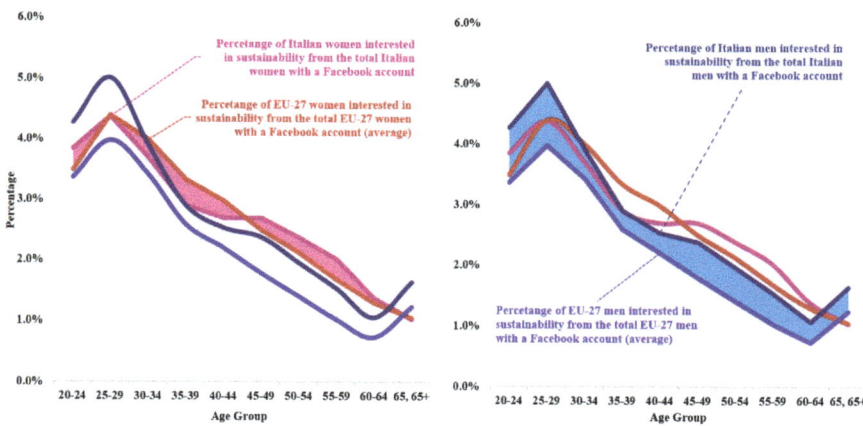

Fig. 15 Italy: Population's interest for sustainability-related subjects. *Source* Authors' development based on Facebook data—Audience Insight (2021)

accounts that also have associated Facebook metadata confirming the interest for the topic of sustainability represent, on average, 2.71% of the Italian men with Facebook accounts (while the EU-27 average was 2.17%).

4.2.16   Latvia

Similar to Bulgaria, Latvia has a performance slightly below the EU-27 average (Fig. 16). As far as the likelihood of Latvian to have a Facebook account is concerned, 33.05% of the Latvian population of age 20 and above is connected to Facebook. More than half (52.48%) of the Latvians connected to the social platforms have also engaged with Facebook posts, pages, events dealing with sustainability-related issues. If reported to the total Latvian population of age 20 and above, only 17.34% of those are connected to Facebook and expressed their concern for sustainability on the social platform.

The percentage of Latvian men with Facebook account who expressed interest in sustainability-related subjects from the total Latvian men that have a Facebook account represented 18.46%, whereas the share of Latvian women with Facebook account, interested in sustainability-related subjects from the total Latvian women that have a Facebook account represented 27.59%, resulting a difference of 0.32% in the favor of the average EU-27 men (reported to Latvian men) and 0.08% in the favor of Latvian women (reported to EU-27 average women).

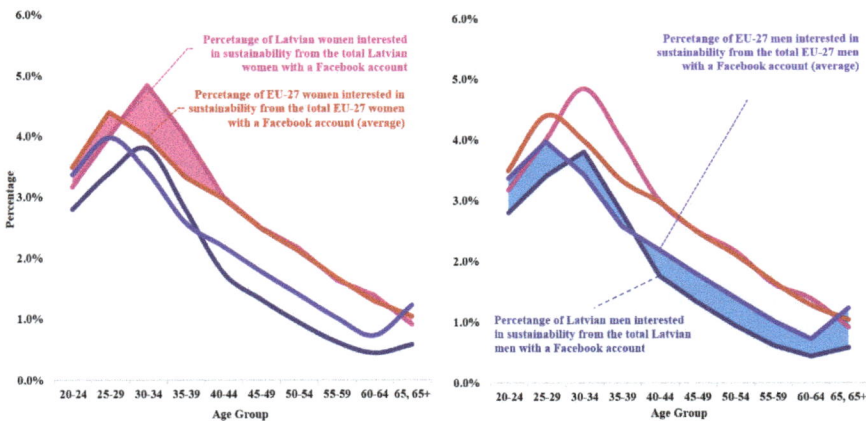

Fig. 16 Latvia: Population's interest for sustainability-related subjects. *Source* Authors' development based on Facebook data—Audience Insight (2021)

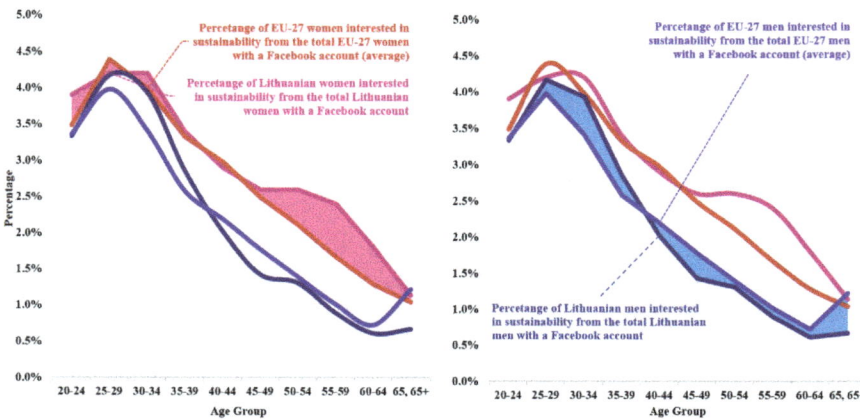

Fig. 17 Lithuania: Population's interest for sustainability-related subjects. *Source* Authors' development based on Facebook data—Audience Insight (2021)

4.2.17 Lithuania

In comparison with Latvia, Lithuania has a slightly performance, even above the EU-27 average in the case of the majority of the age group analyzed (Fig. 17). With a total population of 2,240,420 of age 20 and above (representing only 0.63% of EU-27's population of the same age), 33.05% of those Lithuanian citizens are connected to Facebook and 57.05% of the 33.05% engaged with Facebook pages, events, posts related to sustainability concerns. These statistics show that Lithuanian people fond of technology are more likely to be concerned about sustainability.

A significant behavioral discrepancy was observed between Lithuanian age groups 25–29, 30–34, 35–39, and the rest of the Lithuanian age groups. Similar to the inflection point identified in the case of Germany based on the age factor (Sect. 4.2.10), the Lithuanian inflection point occurs at the age of 39 (5 years older than in the case of Germany). It shows that Lithuanian people tend to engage less on Facebook on sustainability-related topics, posts, and pages after the age of 39, especially in the case of Lithuanian man of age 65 and above.

4.2.18 Luxembourg

Similar to Latvia's situation, Luxembourg has a below-the-average performance when analyzing the civil society's involvement on Facebook and its engagement with sustainability-related posts, events, Facebook pages (Fig. 18). Luxembourg gathers one of the smallest share of EU-27 population of age 20 and above (0.14%), yet only 17.64% of it has engaged with Facebook posts, pages, events related to the topic of sustainability.

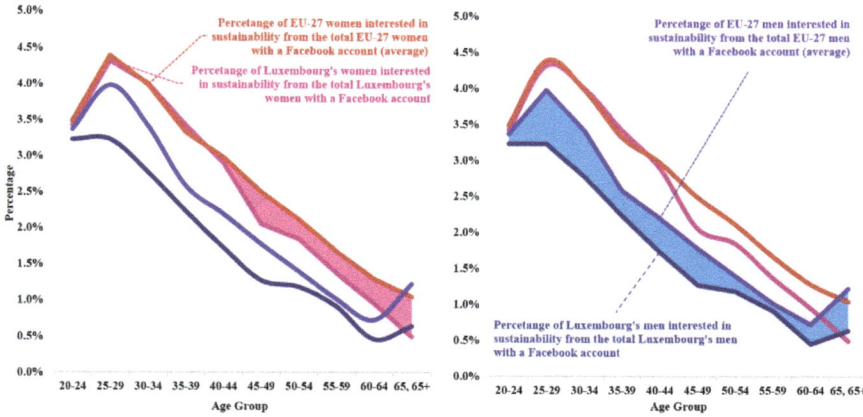

Fig. 18 Luxembourg: Population's interest for sustainability-related subjects. *Source* Authors' development based on Facebook data—Audience Insight (2021)

In the case of Luxembourg men, their engagement with topics specific to sustainability on Facebook is less intense than in the case of women. To support this statement, it is worth mentioning that the average age group percentage of Luxembourg men with Facebook account, interested in sustainability-related subjects from the total Luxembourg men that have a Facebook account represented 1.76%, whereas the average age group percentage of Luxembourg women with Facebook account, interested in sustainability-related subjects from the total Luxembourg women that have a Facebook account represented 2.48%, therefore resulting a difference of 0.72 percentage points in the favor of Luxembourg women.

4.2.19 Malta

Resembling Lithuania's situation (Sect. 4.2.17), Malta has a slightly above-the-average performance than the EU-27 average (Fig. 19). Malta has the smallest share of EU-27's population of age 20 and above: 0.12%, while more than half (51.98%) of Maltese have a Facebook account. The difference between the Maltese citizens of age 20 and above that have Facebook accounts and those that are also expressing their concerns about sustainability on the social network platform is in the favor of the first category: 50.95% vs 20.25%.

The percentage of Maltese men with Facebook accounts who expressed interest in sustainability-related subjects from the total Maltese men that have a Facebook account represented 24.82%, whereas the share of Maltese women with Facebook account, interested in sustainability-related subjects from the total Maltese women that have a Facebook account represented 31.62%, which signals that Maltese women are more concerned about sustainability and expressing it on Facebook much more than Maltese men do.

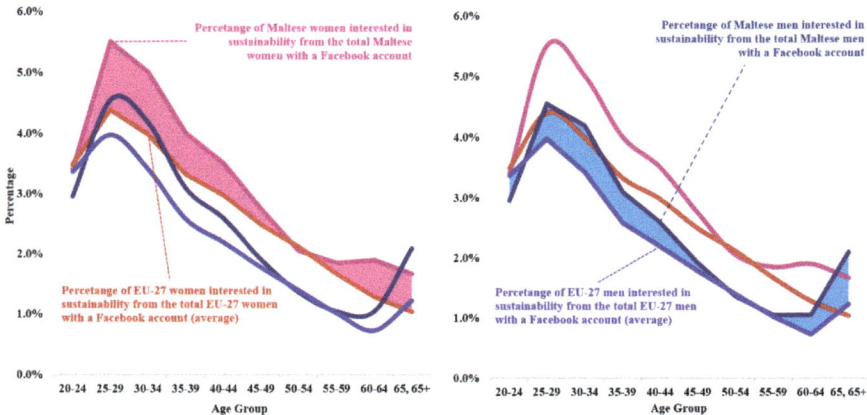

Fig. 19 Malta: Population's interest for sustainability-related subjects. *Source* Authors' development based on Facebook data—Audience Insight (2021)

4.2.20 Netherlands

The Netherlands sums up 3.82% of EU-27's population of age 20 and above, yet less than half of the Dutch people included in this percentage are connected to Facebook (39.61%, very close to the EU-27 average of 39.38%). Situated below the EU-27 average and similar to the case of Luxembourg, the engagement of Dutch men with topics specific to sustainability on Facebook via posts, events, and pages is less intense than in the case of Dutch women (Fig. 20).

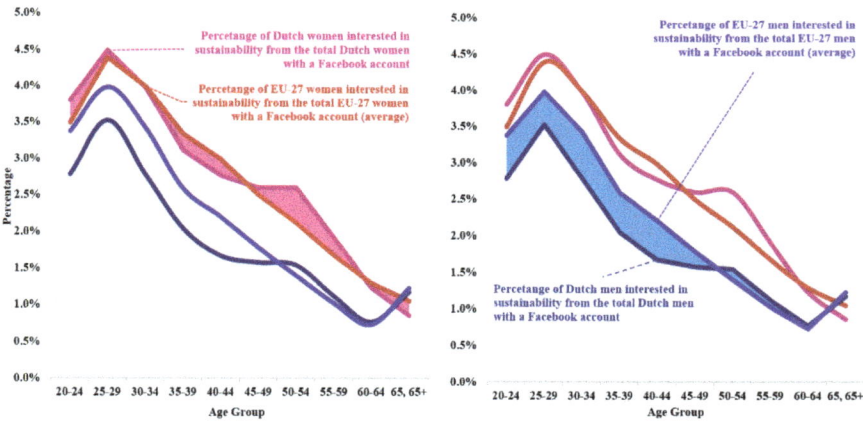

Fig. 20 Netherlands: Population's interest for sustainability-related subjects. *Source* Authors' development based on Facebook data—Audience Insight (2021)

Young Dutch men included in 20–24, 24–29, 30–34, 34–39 age groups differ significantly in digital behavior on Facebook than Dutch women do from the same age groups. In this regard, the average age group percentage of Dutch men (20–39 years old) with Facebook accounts who expressed interest in sustainability-related subjects from the total Dutch men (20–39 years old) linked to Facebook was 2.58%, which represents 1,40 times less than the average corresponding to the same age groups specific to Dutch women.

4.2.21 Poland

Judging through the lens of the methodology of this research, Poland is one of the European countries situated at the bottom of EU-27 ranking based on the percentage of Polish people linked to Facebook and who engage with sustainability-related content on the social network platform (Fig. 21). At the same time, Poland gathers a considerable share of EU-27's population of age 20 and above: 8.50%, out of which only 30.35% have a Facebook account. The difference between the Polish citizens of age 20 and above linked to Facebook and those that are also expressing their concerns about sustainability on the platform is of 15.59%, but not because most of the Facebook-linked citizens people are necessarily engaging with sustainability-related content on Facebook, but because only 22.38% of the Polish population of interest (of age 20 and above) is connected to Facebook (which is the poorest performance in the EU-27).

The average age group percentage of Polish men with Facebook accounts who were interested in sustainability-related subjects from the total Polish men that have a Facebook account represented 0.84%, whereas the average age group percentage of EU-27 men with the same characteristics was 2.17%—therefore resulting a gap

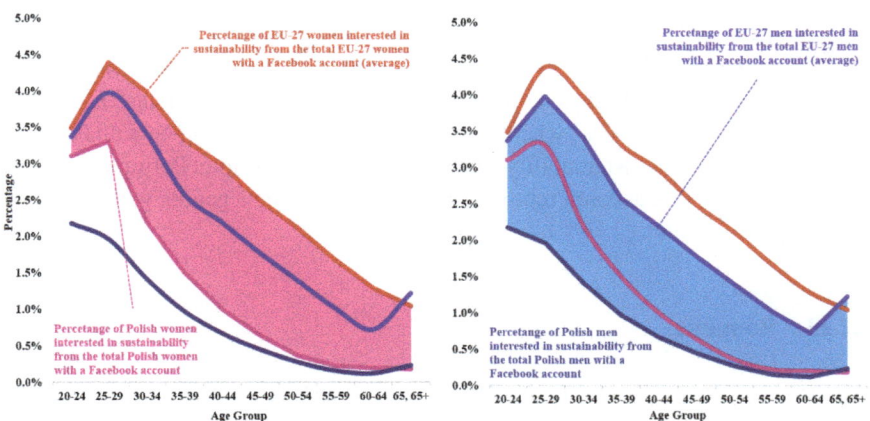

Fig. 21 Poland: Population's interest for sustainability-related subjects. *Source* Authors' development based on Facebook data—Audience Insight (2021)

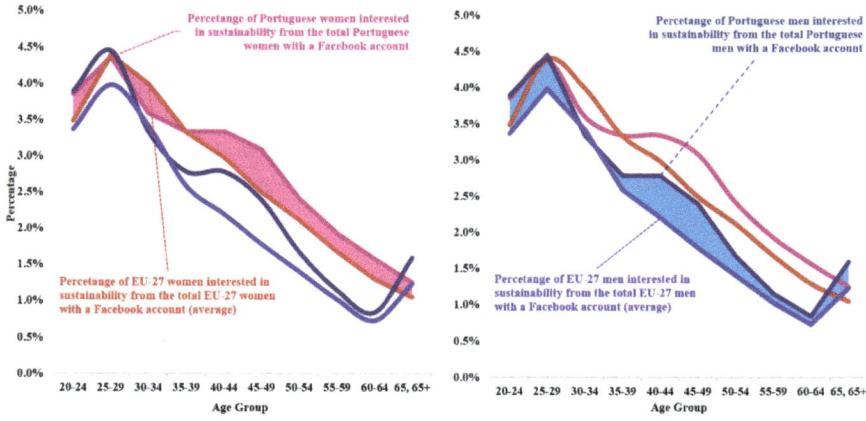

Fig. 22 Portugal: Population's interest for sustainability-related subjects. *Source* Authors' development based on Facebook data—Audience Insight (2021)

of "sustainability interest" of 1.33%. In the case of Polish women, the identified gap was bigger: 1.41%.

4.2.22 Portugal

With a slightly better performance than Malta's, Portugal has the chance to become the EU-27 leaders in the future, based on the percentage of Portuguese people linked to Facebook and who engage with sustainability-related content on the social network platform (Fig. 22). Portugal gathers 2.34% of EU-27's population of age 20 and above, yet less than half (43.10%) have a Facebook account. Out of these 43.10%, more than half (57.14%) engaged with Facebook posts, pages, events related to the topic of sustainable development.

The percentage of Portuguese men with Facebook accounts who expressed interest in sustainability-related subjects from the total Portuguese men that have a Facebook account represented 24.83%, whereas the share of Portuguese women with Facebook account, interested in sustainability-related subjects from the total Portuguese women that have a Facebook account represented 28.62%, causing a gender gap in the favor of Portuguese women (3.79%).

4.2.23 Romania

Romania's performance is similar to that of the Netherland's, based on the percentage of Portuguese people linked to Facebook and who engage with sustainability-related content on the social network (Fig. 23), yet both countries are not in a good spot if compared to the EU-27 average. Romania sums 4.28% of EU-27's population of age

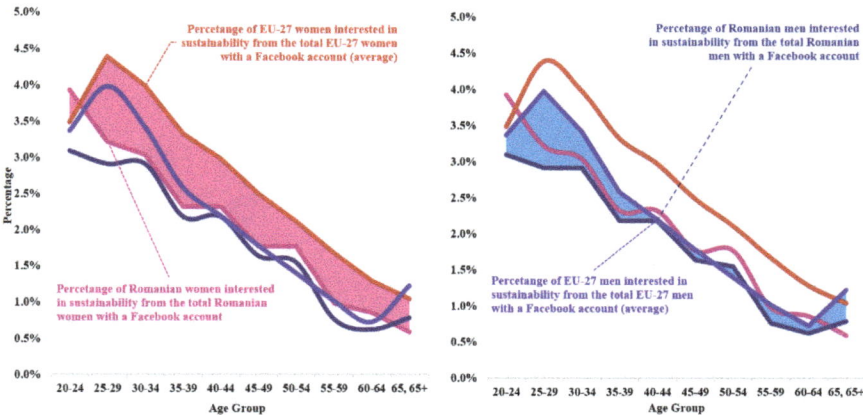

Fig. 23 Romania: Population's interest for sustainability-related subjects. *Source* Authors' development based on Facebook data—Audience Insight (2021)

20 and above, yet few (36.01%) have a Facebook account. Out of the few connected to Facebook, less than half (40.36%) engaged with Facebook posts, pages, events related to the topic of sustainability.

One interesting aspect about the identified Romanian typologies, no matter the gender, is that the average age group percentage differences between the EU-27 and Romanian citizens with Facebook accounts who were also interested in sustainability-related subjects from the total Romanian citizens linked to Facebook was 0.30% (with the same standard deviation) in the case of man and double in the case of women (with a smaller standard deviation: 0.43%).

4.2.24 Slovakia

According to the research methodology, together with Poland, Slovakia is a country situated at the bottom of the EU-27 ranking based on the percentage of citizens linked to Facebook and who engage with sustainability-related content on the social network platform (Fig. 24). Slovakia gathers a small share of EU-27's population of age 20 and above: 1.21%, out of which only 32.32% have a Facebook account. Only 14.20% Slovakian citizens are interested in Facebook posts, pages, and events related to sustainability out of the 4,331,402 Slovakian citizens of age 20 and above.

The average age group percentage of Slovakian men with Facebook accounts who were interested in sustainability-related subjects from the total Slovakian men that have a Facebook account represented 1.79%, whereas the average age group percentage of EU-27 men with the same characteristics was 2.17%—therefore resulting a gap of "sustainability interest" of 0.38% (better than in the case of Polish men). Regarding Slovakian women, the identified gap was 0.46%.

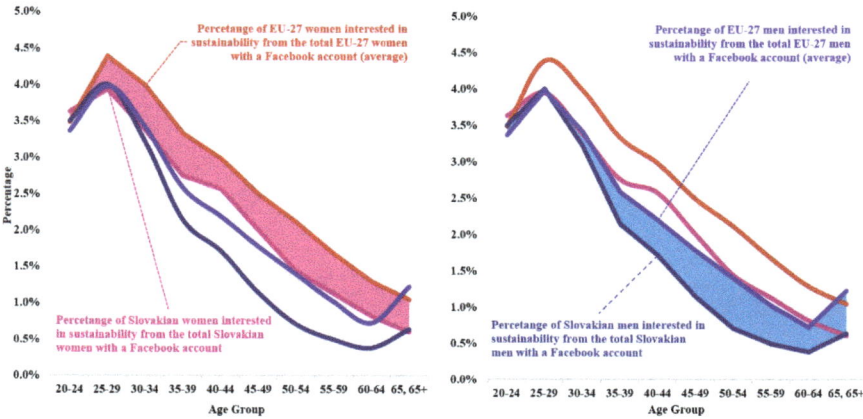

Fig. 24 Slovakia: Population's interest for sustainability-related subjects. *Source* Authors' development based on Facebook data—Audience Insight (2021)

4.2.25   Slovenia

Almost identical with Slovakia's performance, Slovenia is another below-the-average performer in the EU-27, based on the percentage of citizens linked to Facebook and who engage with sustainability-related content on the social network platform (Fig. 25). Slovenia gathers an ever smaller share of EU-27's population of age 20 and above: 0.47%, out of which 34.39% are linked to Facebook. Only 14.05% Slovenian citizens are interested in Facebook posts, pages, and events related to sustainability out of the 1,686,330 Slovenian citizens of age 20 and above.

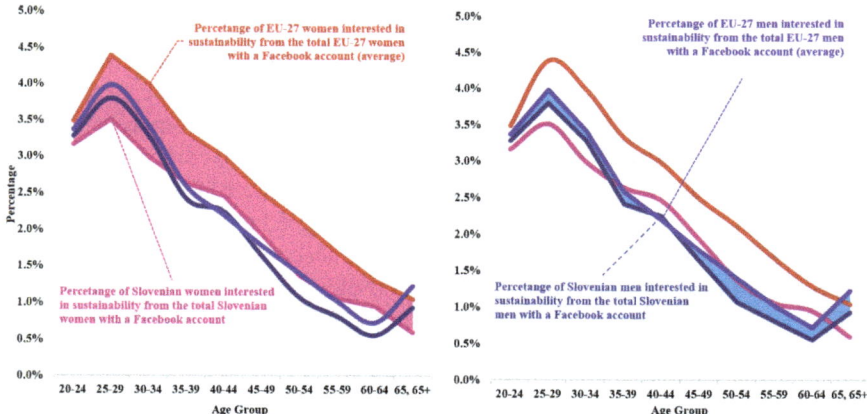

Fig. 25 Slovenia: Population's interest for sustainability-related subjects. *Source* Authors' development based on Facebook data—Audience Insight (2021)

The percentage of Slovenian men with Facebook accounts who expressed interest in sustainability-related subjects from the total Slovenian men that linked to Facebook represented 20.00%, whereas the share of Slovenian women with Facebook account, interested in sustainability-related subjects from the total Slovenian women that connected to a Facebook account represented 20.63%, causing a very small gender gap to occur in the favor of Slovenian women (0.63%). This finding shows little behavioral differences on Facebook based on gender in the case of Slovenia.

4.2.26 Spain

Among all of the analyzed age groups, Spanish men between 25 and 29 years old are the most interested age group about sustainability-related actions, events, subjects, according to Facebook Audience Insight (Fig. 26). If compared to Spanish women, referring to the same category (i.e.: 25–29 years old men), Spanish men are more likely to be advocating for sustainability, considering that the percentage of Spanish men with Facebook accounts interested in sustainability-related subjects from the total Spanish men connected to Facebook represented 4.73%, whereas the percentage reached 3.81% in the case of Spanish women (identified gender gap: 0.92%).

Spanish people of age 20 and above represent 10.67% of EU-27's population within the same age group. However, less than 40% of Spanish people are connected to Facebook. The average age group percentage of Spanish men with Facebook accounts who were interested in sustainability-related subjects from the total Spanish men connected to Facebook was 2.25%, whereas the average age group percentage of EU-27 men with the same characteristics was 2.17%, signaling a better performance than the average in the case of Spain. However, not the same positive result was observed when the average age group percentage of Spanish women was analyzed

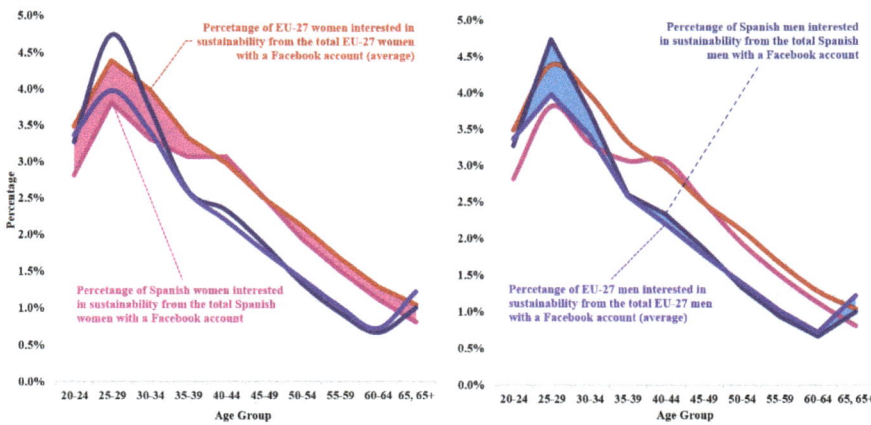

Fig. 26 Spain: Population's interest for sustainability-related subjects. *Source* Authors' development based on Facebook data—Audience Insight (2021)

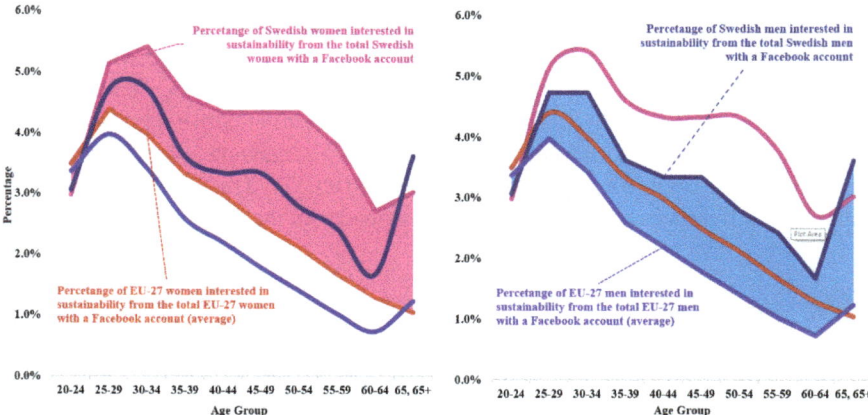

Fig. 27 Sweden: Population's interest for sustainability-related subjects. *Source* Authors' development based on Facebook data—Audience Insight (2021)

in relation with the EU-27 average (identified "interest" gap: 0.29% in the favor of EU-27).

4.2.27 Sweden

One of the top three EU-27 leaders when analyzing the percentage of citizens expressing their preferences toward sustainability on Facebook (Fig. 27), Sweden has a population of 7,923,859 people of age 20 and above (2.22% of EU-27's population of the same age). An important aspect specific to Sweden's performance is that 77.97% of the Swedish citizens of age 20 and above who are connected to Facebook have also engaged with Facebook posts, events, and pages that were related to sustainability-related issues, whereas the EU-27 average was 50.95%.

The percentage of Swedish men with Facebook accounts who expressed interest in sustainability-related subjects from the total Swedish men that linked to Facebook represented 33.25%, whereas the share of Swedish women with Facebook account, interested in sustainability-related subjects from the total Swedish women that connected to a Facebook account represented 40.58%, whereas the EU-27 average was 21.67% in the case of men and 26.76% in the case of women. These behavioral differences (men: 11.58% and women: 13.82%) in the direction of empowering sustainability on Facebook place Sweden in the EU-27 ranking.

4.3 Econometric Models Predicting the Level of Sustainable Development Based on Facebook Data

Multiple cross-sectional linear regression models were constructed with the purpose of trying to predict the level of sustainable development in the EU-27 by resorting to the data extracted from Facebook Audience Insight. All the models consisted of only two variables and the SDG Index always played the role of the dependent variable, as stated in Table 4.

The greatest values of the coefficient of determination (R^2) was registered in the case of I_9 (0.344), the percentage population with Facebook account that is interested in sustainability-related subjects from the national population. This means that, in the EU-27, 34.4% of the variation of the SDG Index is explained by the variation of the percentage population with Facebook account that is interested in sustainability-related subjects from the national population. These results are valid considering the F-statistics value (9.896) and its corresponding p-value (0.004), below the 0.05 threshold. The coefficient of I_9 predictor shows that each percentage of the population with Facebook account that is interested in sustainability-related subjects from the national population has a corresponding multiplier of 19.755, to which must be added the constant of 72.691. The p-values corresponding to the constant and to the coefficient of I_9 predictor are below the 0.05 threshold, validating the constructed model.

Besides the econometric results, one conclusion that can be drawn from this model is that EU-27 citizens engaging with sustainability-related posts, Facebook pages, and events can only partially explain the level of sustainable development of the EU-27 countries. Causes are multiple: (a) some countries have already reached high levels of sustainable development (examples: Sweden, Denmark, Finland), yet citizens continue to be eager to engage with related content on Facebook, monitor progress and take action in the direction of change (see results from Sects. 4.2.7, 4.2.9 and 4.2.27); (b) other countries that are currently trying to improve their performance in relation with the SDGs and their specific targets (examples: Bulgaria, Romania, Cyprus), while citizens are not as motivated to look up sustainability-related initiatives on Facebook (see results from Sects. 4.2.3, 4.2.5 and 4.2.23) as the citizens from countries performing well in relation with achieving in the SDGs in a timely manner; (c) the cultural background (DeAndrea et al., 2010) and level of digital infrastructures (Tilson et al., 2010) might influence the predisposition of citizens to engage on platforms like Facebook.

Regarding the developed typologies of EU-27 citizens and their corresponding, they were integrated in the econometric model by resorting to the following indicator: *Percetange of population interested in sustainability on Facebook from the total population of age 20 and above who own a Facebook account*, per age group and gender (EU-27 average, existing 27 observations per age group). In the model, the previously mentioned indicator played the role of the independent variable, while the SDG Index was dependent. The greatest success in predicting the variance of the SDG Index was observed in relation with the predictor specific to the 25–29 age

Table 4 Categories of food and agricultural products included in the study

Formula of the equation: SDG Index = Constant + Coefficient of Predictor × Dependent Variable + ε

Dependent Variables			Constant	Coefficient of Predictor	R^2	Adjusted R^2	Durbin Watson	F-statistics	p-value of F-statistics
Percentage of population interested in sustainability on Facebook from the total population of age 20 and above who own a Facebook account	20–24	M	74.067	134.495*	0.064	0.026	1.167	1.701	0.204
		W	72.141	179.234*	0.097	0.061	0.984	2.685	0.114
	25–29	M	72.680	148.756	0.137	0.102	1.328	3.963	0.058
		W	69.406	209.314	0.318	0.291	1.080	11.656	0.002
	30–34	M	73.649	144.803*	0.121	0.086	1.315	3.449	0.075
		W	71.831	169.747	0.242	0.211	1.175	7.962	0.009
	35–39	M	74.963	140.594*	0.090	0.053	1.335	2.459	0.129
		W	73.447	154.659	0.193	0.161	1.285	5.978	0.022
	40–44	M	75.640	134.552*	0.081	0.044	1.314	2.203	0.150
		W	74.383	141.454	0.159	0.125	1.293	4.721	0.039
	45–49	M	75.493	174.296	0.146	0.112	1.402	4.278	0.049
		W	74.476	165.248	0.231	0.200	1.308	7.490	0.011
	50–54	M	75.931	190.406	0.135	0.101	1.402	3.918	0.059
		W	75.180	161.353	0.247	0.217	1.361	8.186	0.008
	55–59	M	75.915	263.382	0.212	0.180	1.458	6.708	0.016
		W	75.706	172.595	0.270	0.240	1.382	9.230	0.006
	60–64	M	76.344	309.809	0.161	0.128	1.449	4.803	0.038
		W	76.212	184.948	0.175	0.142	1.352	5.286	0.030
	65 and above	M	76.366	181.883	0.252	0.222	1.461	8.437	0.008
		W	76.339	216.546	0.289	0.261	1.434	10.167	0.004

(continued)

Table 4 (continued)

Formula of the equation: SDG Index = Constant + Coefficient of Predictor × Dependent Variable + ε

Dependent Variables	Constant	Coefficient of Predictor	R^2	Adjusted R^2	Durbin Watson	F-statistics	p-value of F-statistics
I_8	72.691	11.582	0.284	0.255	1.368	9.896	0.004
I_9	72.528	19.755	0.344	0.318	1.193	13.117	0.001

*Signals a corresponding p-value above the 0.05 accepted threshold. If not marked, the corresponding p-value of the Constant and of the Coefficient of Predictor is below the 0.05 threshold

Abbreviation: "M" stands for mEn and "W" stands for women

Source Authors' development based on Facebook, SDG andEurostat (2021)

group of female gender, with a corresponding coefficient of determination of 0.318. This means that 31.8% of the SDG Index variance in the EU-27 is explained by the variance of the percentage of EU-27 women of age 25–29 who are interested in sustainability-related issues on Facebook from the total EU-27 population of age 20 and above who are connected to Facebook via an account. These results are valid: (a) the F-statistics value (11.655) and its corresponding p-value (0.002), below the 0.05 threshold and (b) the p-values corresponding to the constant and to the coefficient of predictor are below the 0.05 threshold as well. The coefficient of predictor shows that each *percentage of 20–24 years old EU-27 women interested in sustainability on Facebook from the total EU-27 women population of age 20 and above who own a Facebook account* has a corresponding multiplier of 209.313, to which must be added the constant of 72.527. Although the R^2 and Adjusted R^2 values are relatively low, this does not mean the constructed econometric model is irrelevant, but rather it means that Facebook data cannot successfully predict the actual level of sustainable development of the EU-27 members, because the human factor and digital behaviors are not convergent with the level of sustainable development: highly performant countries in this regard have an active civil society on the social network platform, while the civil society from less-performant countries are not that eager to engage on Facebook with sustainability-related content. Causes are multiple: (a) the level of development and quality of life play a huge part in culture and shape the nature and intensity of day-to-day concerns (Ignat & Constantin, 2020; Pătărlăgeanu et al., 2020); (b) the digital infrastructure is lacking in some EU-27 countries (Butenko et al., 2021); (c) social network platform continue to evolve and Facebook is not the only platform where European citizens can engage with sustainability-related content (Huang et al., 2019; Mamonov et al., 2016).

5 Conclusions

Achieving higher levels of sustainable development is tied to the constant process of transformation, which also involves the digital component. Understanding the impact of digital transformation on becoming more sustainable involves resorting to a valuable asset, which is data. One of the greatest predictor of behavioral change toward the natural capital, society and economy is personal data. In this regard, Facebook is one of the biggest digital platforms that can provide intel on paradigm changes specific to different typologies of people.

This book chapter aimed at filling a literature gap regarding the possibilities of predicting patterns in the direction of sustainable development by resorting to big data management. With the Facebook Audience Insight collected data, multiple digital profiles of EU-27 citizens were designed based on: (i) their predisposition to engage with Facebook sustainability-related content; (ii) the age factor; (iii) the gender factor. In this research, attention was paid to the relation between the level of sustainable development in the EU-27 and the interest of citizens for the content existing on Facebook dealing with the topic of sustainability.

Answering to the research question that represents the title of this book chapter was made possible by constructing and testing multiple econometric models designed to predict the level of sustainable development in the EU-27 by resorting to SDG Index and Facebook Audience Insight data. Results showed that analyzing different typologies of EU-27 citizens and their interaction with sustainability-related content on Facebook is not sufficient to successfully predict the actual level of sustainable development of the countries corresponding to the typologies of analyzed citizens. These findings provide an answer much complex than a simple "yes or no" type of answer. The level of sustainable development in the EU-27 is only partially explained by the online behavior of the EU-27 citizens connected to Facebook and the reasons for this are numerous: the countries with high levels of sustainable development (Sweden, Denmark, Finland) have a very active civil society that advocate for sustainability on the social network, while the civil society from other countries less performant from a sustainable development perspective are not eager to engage with content specific to sustainability on Facebook. Reasons are multiple: different cultural background, lack of digital infrastructure, and others.

There are several limitations attached to this research: (1) the interest for sustainability-related content on Facebook is different from actually taking action in real life in this direction: one might advocate for sustainable development on the social network, yet do the opposite through its day-to-day decisions; (2) the SDG Index is a composite that expresses the level of sustainable development through the lens of selected indicators, as defined in the methodology of Sachs et al. (2020); (3) Facebook might not be the most popular social network platform in all the EU-27 countries, some other online platforms could be of much more interest.

Future research avenues might be expanding the analysis with other typologies of citizens, as means to add more variables in the econometric models designed to predict the level of sustainable development based on Facebook-collected data. In this regard, more research queries can be run in order to collect and process data from Facebook Audience Insight platform. By doing that, more behavioral patterns would be explored and used as predictor variables of sustainable development.

References

Ajana, B. (2020). Personal metrics: Users' experiences and perceptions of self-tracking practices and data. *Social Science Information, 59*(4), 654–678. https://doi.org/10.1177/053901842095 9522

Ali, S. H., Foreman, J., Capasso, A., Jones, A. M., Tozan, Y., & DiClemente, R. J. (2020). Social media as a recruitment platform for a nationwide online survey of COVID-19 knowledge, beliefs, and practices in the United States: Methodology and feasibility analysis. *BMC Medical Research Methodology, 20*(1), 116. https://doi.org/10.1186/s12874-020-01011-0

Al-Jarrah, O. Y., Yoo, P. D., Muhaidat, S., Karagiannidis, G. K., & Taha, K. (2015). Efficient machine learning for big data: A review. *Big Data Research, 2*(3), 87–93. https://doi.org/10.1016/j.bdr. 2015.04.001

Andrei, J. V., Constantin, M., & de los Ríos Carmenado, I. (2021). Assessing EU's progress and performance with regard to SDG-12 targets and indicators. In C. J. Chiappetta Jabbour & S. A. R. Khan (Eds.), *Sustainable production and consumption systems*. Springer.

Añón Higón, D., Gholami, R., & Shirazi, F. (2017). ICT and environmental sustainability: A global perspective. *Telematics and Informatics, 34*(4), 85–95. https://doi.org/10.1016/j.tele.2017.01.001

Aysan, A. F., Bergigui, F., & Disli, M. (2021). Using blockchain-enabled solutions as SDG accelerators in the international development space. *Sustainability, 13*(7), 4025. https://doi.org/10.3390/su13074025

Baronian, L. (2020). Digital platforms and the nature of the firm. *Journal of Economic Issues, 54*(1), 214–232. https://doi.org/10.1080/00213624.2020.1720588

Barr, S. (2003). Strategies for sustainability: Citizens and responsible environmental behaviour. *Area, 35*(3), 227–240.

Beier, G., Niehoff, S., & Xue, B. (2018). More sustainability in industry through industrial internet of things? *Applied Sciences, 8*(2), 219. https://doi.org/10.3390/app8020219

Beier, G., Ullrich, A., Niehoff, S., Reißig, M., & Habich, M. (2020). Industry 4.0: How it is defined from a sociotechnical perspective and how much sustainability it includes—A literature review. *Journal of Cleaner Production, 259*, 120856. https://doi.org/10.1016/j.jclepro.2020.120856

Bifulco, F., Tregua, M., Amitrano, C. C., & D'Auria, A. (2016). ICT and sustainability in smart cities management. *International Journal of Public Sector Management, 29*(2), 132–147. https://doi.org/10.1108/IJPSM-07-2015-0132

Birch, K., Cochrane, D., & Ward, C. (2021). Data as asset? The measurement, governance, and valuation of digital personal data by Big Tech. *Big Data & Society, 8*(1), 20539517211017308. https://doi.org/10.1177/20539517211017308

Bradley, K. (2007). Defining digital sustainability. *Library Trends, 56*(1), 148–163. https://doi.org/10.1353/lib.2007.0044

Bucher, T. (2012). Want to be on the top? Algorithmic power and the threat of invisibility on Facebook. *New Media & Society, 14*(7), 1164–1180. https://doi.org/10.1177/1461444812440159

Butenko, E. D., Korolev, V. A., Peredereeva, E. V., & Isakhaev, N. R. (2021). Digital infrastructure development in Russia, Europe and Asia. In E. G. Popkova & B. S. Sergi (Eds.), *Modern global economic system: Evolutional development vs. revolutionary leap* (pp. 1493–1502). Springer International Publishing. https://doi.org/10.1007/978-3-030-69415-9_165

Calvo, D., Cano-Orón, L., & Baviera, T. (2021). Global spaces for local politics: An exploratory analysis of Facebook Ads in Spanish election campaigns. *Social Sciences, 10*(7), 271. https://doi.org/10.3390/socsci10070271

Celebi, S. I. (2015). How do motives affect attitudes and behaviors toward internet advertising and Facebook advertising? *Computers in Human Behavior, 51*, 312–324. https://doi.org/10.1016/j.chb.2015.05.011

Constantin, M., Dinu, M., Pătărlăgeanu, S. R., & Chelariu, C. (2021). Sustainable development disparities in the EU-27 based on R&D and innovation factors. *Amfiteatru Economic, 26*, 672–687.

Constantin, M., Pătărlăgeanu, S. R., Dinu, M., & Ignat, R. (2021). Rising tensions along the agri-food value chains during the COVID-19 crisis: Evidence based on google trends data. *Proceedings of the International Conference on Business Excellence, 15*, 302–316. https://doi.org/10.2478/picbe-2021-0029

Constantin, M., Rădulescu, I. D., Vasić, M., Andrei, J. V., & Pistalu, M. (2021). Modern forms of value creation in the global agri-food chain–perspectives from the facebook ads market. *Proceedings of the International Conference on Business Excellence, 15*, 811–823. https://doi.org/10.2478/picbe-2021-0075

Constantin, M., Strat, G., Deaconu, M. E., & Pătărlăgeanu, S. R. (2021). Innovative agri-food value chain management through a unique urban ecosystem. *Management Research and Practice, 13*(3), 5–22.

Davies, A. R., & Legg, R. (2018). Fare sharing: Interrogating the nexus of ICT, urban food sharing, and sustainability. *Food, Culture & Society, 21*(2), 233–254. https://doi.org/10.1080/15528014.2018.1427924

DeAndrea, D. C., Shaw, A. S., & Levine, T. R. (2010). Online language: The role of culture in self-expression and self-construal on Facebook. *Journal of Language and Social Psychology, 29*(4), 425–442. https://doi.org/10.1177/0261927X10377989

De Bernardi, P., Bertello, A., & Venuti, F. (2019). Online and on-site interactions within alternative food networks: Sustainability impact of knowledge-sharing practices. *Sustainability, 11*(5), 1457. https://doi.org/10.3390/su11051457

Dehghani, M., & Tumer, M. (2015). A research on effectiveness of Facebook advertising on enhancing purchase intention of consumers. *Computers in Human Behavior, 49*, 597–600. https://doi.org/10.1016/j.chb.2015.03.051

Del Río Castro, G., González Fernández, M. C., & Uruburu Colsa, Á. (2021). Unleashing the convergence amid digitalization and sustainability towards pursuing the Sustainable Development Goals (SDGs): A holistic review. *Journal of Cleaner Production, 280*, 122204. https://doi.org/10.1016/j.jclepro.2020.122204

Diaz-Sarachaga, J. M., Jato-Espino, D., & Castro-Fresno, D. (2018). Is the Sustainable Development Goals (SDG) index an adequate framework to measure the progress of the 2030 Agenda? *Sustainable Development, 26*(6), 663–671. https://doi.org/10.1002/sd.1735

Di Vaio, A., Palladino, R., Hassan, R., & Escobar, O. (2020). Artificial intelligence and business models in the sustainable development goals perspective: A systematic literature review. *Journal of Business Research, 121*, 283–314. https://doi.org/10.1016/j.jbusres.2020.08.019

Duch-Brown, N., & Rossetti, F. (2020). Digital platforms across the European regional energy markets. *Energy Policy, 144*, 111612. https://doi.org/10.1016/j.enpol.2020.111612

Enli, G. S., & Skogerbø, E. (2013). Personalized campaigns in party-centred politics. *Information, Communication & Society, 16*(5), 757–774. https://doi.org/10.1080/1369118X.2013.782330

Eurostat. (2021). *Eurostat Database Online.* https://ec.europa.eu/eurostat/databrowser/view/demo_pjangroup/default/table?lang=en. Accessed June 2021.

Facebook. (2021a). *Facebook Audience Insight Database.* https://www.facebook.com/business/insights/tools/audience-insights. Accessed June 2021.

Facebook. (2021b). *Facebook Reports Fourth Quarter and Full Year 2020 Results.*

Fernandez, K. C., Levinson, C. A., & Rodebaugh, T. L. (2012). Profiling: Predicting social anxiety from Facebook profiles. *Social Psychological and Personality Science, 3*(6), 706–713. https://doi.org/10.1177/1948550611434967

Gavino, M. C., Williams, D. E., Jacobson, D., & Smith, I. (2018). Latino entrepreneurs and social media adoption: Personal and business social network platforms. *Management Research Review, 42*(4), 469–494. https://doi.org/10.1108/MRR-02-2018-0095

Gonzalez-Lafaysse, L., & Lapassouse-Madrid, C. (2016). Facebook and sustainable development: A case study of a French supermarket chain. *International Journal of Retail & Distribution Management, 44*(5), 560–582. https://doi.org/10.1108/IJRDM-01-2015-0012

Goodenough, D. G., Bhogal, P., Charlebois, D., Matwin, S., & Niemann, O. (1995). Intelligent data fusion for environmental monitoring. *1995 International Geoscience and Remote Sensing Symposium, IGARSS '95. Quantitative Remote Sensing for Science and Applications, 3*, 2157–2160. https://doi.org/10.1109/IGARSS.1995.524135

Goodenough, D. G., Charlebois, D., Bhogal, A. S., & Daley, N. (1998). An improved planner for intelligent monitoring of sustainable development of forests. *IGARSS '98. Sensing and Managing the Environment. 1998 IEEE International Geoscience and Remote Sensing. Symposium Proceedings. (Cat. No.98CH36174), 1*, 397–399. https://doi.org/10.1109/IGARSS.1998.702917

Gouvea, R., Kapelianis, D., & Kassicieh, S. (2018). Assessing the nexus of sustainability and information & communications technology. *Technological Forecasting and Social Change, 130*, 39–44. https://doi.org/10.1016/j.techfore.2017.07.023

Günther, W. A., Rezazade Mehrizi, M. H., Huysman, M., & Feldberg, F. (2017). Debating big data: A literature review on realizing value from big data. *The Journal of Strategic Information Systems, 26*(3), 191–209. https://doi.org/10.1016/j.jsis.2017.07.003

Hilty, L. M., & Aebischer, B. (2015). ICT for sustainability: An emerging research field. In L. M. Hilty & B. Aebischer (Eds.), *ICT innovations for sustainability* (pp. 3–36). Springer International Publishing. https://doi.org/10.1007/978-3-319-09228-7_1

How, M.-L., Cheah, S.-M., Chan, Y.-J., Khor, A. C., & Say, E. M. P. (2020). Artificial Intelligence-enhanced decision support for informing global sustainable development: A human-centric AI-thinking approach. *Information, 11*(1), 39. https://doi.org/10.3390/info11010039

Huang, L., Clarke, A., Heldsinger, N., & Tian, W. (2019). The communication role of social media in social marketing: A study of the community sustainability knowledge dissemination on LinkedIn and Twitter. *Journal of Marketing Analytics, 7*(2), 64–75. https://doi.org/10.1057/s41270-019-00053-8

Ignat, R., & Constantin, M. (2020). Multidimensional facets of entrepreneurial resilience during the COVID-19 crisis through the lens of the Wealthiest Romanian Counties. *Sustainability, 12*(23), 10220. https://doi.org/10.3390/su122310220

Khan, F. N., Sana, A., & Arif, U. (2020). Information and communication technology (ICT) and environmental sustainability: A panel data analysis. *Environmental Science and Pollution Research, 27*(29), 36718–36731. https://doi.org/10.1007/s11356-020-09704-1

Kim, C., & Shen, C. (2020). Connecting activities on Social Network Sites and life satisfaction: A comparison of older and younger users. *Computers in Human Behavior, 105*, 106222. https://doi.org/10.1016/j.chb.2019.106222

Koïvogui, A., Levi, S., Finkler, M., Lewkowicz, S., Gombeaud, T., Sabate, J. M., Duclos, C., & Benamouzig, R. (2020). Feasibility of encouraging participation in colorectal cancer screening campaigns by motivating people through the social network, Facebook. *Colorectal Disease: the Official Journal of the Association of Coloproctology of Great Britain and Ireland, 22*(10), 1325–1335. https://doi.org/10.1111/codi.15121

König, R., Uphues, S., Vogt, V., & Kolany-Raiser, B. (2020). The tracked society: Interdisciplinary approaches on online tracking. *New Media & Society, 22*(11), 1945–1956. https://doi.org/10.1177/1461444820924629

Leung, C. K., Jiang, F., Poon, T. W., & Crevier, P.-É. (2018). Big data analytics of social network data: Who cares most about you on Facebook? In M. Moshirpour, B. Far, & R. Alhajj (Eds.), *Highlighting the importance of Big Data management and analysis for various applications* (pp. 1–15). Springer International Publishing. https://doi.org/10.1007/978-3-319-60255-4_1

Liu, P., Bai, Y., Wang, L., & Li, X. (2017). Partial k-Anonymity for privacy-preserving social network data publishing. *International Journal of Software Engineering and Knowledge Engineering, 27*(01), 71–90. https://doi.org/10.1142/S0218194017500048

MacFeely, S. (2019). The Big (data) Bang: Opportunities and challenges for compiling SDG indicators. *Global Policy, 10*(S1), 121–133. https://doi.org/10.1111/1758-5899.12595

Mamonov, S., Koufaris, M., & Benbunan-Fich, R. (2016). The role of the sense of community in the sustainability of social network sites. *International Journal of Electronic Commerce, 20*(4), 470–498. https://doi.org/10.1080/10864415.2016.1171974

Margherita, A., & Heikkilä, M. (2021). Business continuity in the COVID-19 emergency: A framework of actions undertaken by world-leading companies. *Business Horizons, 64*(5), 683–695. https://doi.org/10.1016/j.bushor.2021.02.020

Melanthiou, Y., Pavlou, F., & Constantinou, E. (2015). The use of social network sites as an E-Recruitment tool. *Journal of Transnational Management, 20*(1), 31–49. https://doi.org/10.1080/15475778.2015.998141

Nañez Alonso, S. L., Reier Forradellas, R. F., Pi Morell, O., & Jorge-Vazquez, J. (2021). Digitalization, circular economy and environmental sustainability: The application of artificial intelligence in the efficient self-management of waste. *Sustainability, 13*(4), 2092. https://doi.org/10.3390/su13042092

N'dri, L. M., Islam, M., & Kakinaka, M. (2021). ICT and environmental sustainability: Any differences in developing countries? *Journal of Cleaner Production, 297*, 126642. https://doi.org/10.1016/j.jclepro.2021.126642

Nigro, M., Ferrara, M., De Vincentis, R., Liberto, C., & Valenti, G. (2021). Data driven approaches for sustainable development of E-Mobility in Urban Areas. *Energies, 14*(13), 3949. https://doi.org/10.3390/en14133949

Nolin, J. M. (2019). Data as oil, infrastructure or asset? Three metaphors of data as economic value. *Journal of Information, Communication and Ethics in Society, 18*(1), 28–43. https://doi.org/10.1108/JICES-04-2019-0044

Nuccio, M., & Guerzoni, M. (2019). Big data: Hell or heaven? Digital platforms and market power in the data-driven economy. *Competition & Change, 23*(3), 312–328. https://doi.org/10.1177/1024529418816525

Ordieres-Meré, J., Prieto Remón, T., & Rubio, J. (2020). Digitalization: An opportunity for contributing to sustainability from knowledge creation. *Sustainability, 12*(4), 1460. https://doi.org/10.3390/su12041460

Pallas, J., Eidenfalk, J., & Engel, S. (2019). Social networking sites and learning in international relations: The impact of platforms. *Australasian Journal of Educational Technology, 35*(1), Article 1. https://doi.org/10.14742/ajet.3637

Pătărlăgeanu, S. R., Rădulescu, C. V., Dinu, M., & Constantin, M. (2020). The impact of heavy work investment on the economy and the individual. *Amfiteatru Economic, 22*(SI 14), 1085–1102.

Petrariu, R., Constantin, M., Dinu, M., Pătărlăgeanu, S. R., & Deaconu, M. E. (2021). Water, energy, food, waste nexus: Between synergy and trade-offs in Romania based on entrepreneurship and economic performance. *Energies, 14*(16), 5172. https://doi.org/10.3390/en14165172

Popescu, M.-F., Chiripuci, B.-C., Orîndaru, A., Constantin, M., & Scrieciu, A. (2020). Fostering sustainable development through shifting toward rural areas and digitalization—The case of Romanian Universities. *Sustainability, 12*(10), 4020. https://doi.org/10.3390/su12104020

Pöyry, E., Parvinen, P., & Malmivaara, T. (2013). Can we get from liking to buying? Behavioral differences in hedonic and utilitarian Facebook usage. *Electronic Commerce Research and Applications, 12*(4), 224–235. https://doi.org/10.1016/j.elerap.2013.01.003

Rhum, K. (2021). Information fiduciaries and political microtargeting: A legal framework for regulating political advertising on digital platforms. *Northwestern University Law Review, 115*(6), 1829–1874.

Robèrt, K.-H., Daly, H., Hawken, P., & Holmberg, J. (1997). A compass for sustainable development. *International Journal of Sustainable Development & World Ecology, 4*(2), 79–92. https://doi.org/10.1080/13504509709469945

Roblek, V., Meško, M., Bach, M. P., Thorpe, O., & Šprajc, P. (2020). The interaction between internet, sustainable development, and emergence of society 5.0. *Data, 5*(3), 80. https://doi.org/10.3390/data5030080

Sabin, J., & Olive, A. (2018). Slack: Adopting social-networking platforms for active learning. *PS: Political Science & Politics, 51*(1), 183–189. https://doi.org/10.1017/S1049096517001913

Sacală, M.-D., Pătărlăgeanu, S. R., Popescu, M.-F., & Constantin, M. (2021). Econometric research on the mix of factors influencing first-year students' dropout decision at the Faculty of agri-food and environmental economics. *Economic Computation & Economic Cybernetics Studies & Research, 55*(3).

Sachs, J., Schmidt-Traub, G., Kroll, C., Lafortune, G., Fuller, G., & Woelm, F. (2020). *Sustainable Development Report 2020*. Cambridge University Press.

Sadowski, J. (2020). The Internet of Landlords: Digital Platforms and New Mechanisms of Rentier Capitalism. *Antipode, 52*(2), 562–580. https://doi.org/10.1111/anti.12595

Shao, W., Ross, M., & Grace, D. (2015). Developing a motivation-based segmentation typology of Facebook users. *Marketing Intelligence & Planning, 33*(7), 1071–1086. https://doi.org/10.1108/MIP-01-2014-0014

Shih, C. (2010). *The Facebook Era: Tapping Online social networks to market, sell, and innovate*. Pearson Education.

Sivarajah, U., Kamal, M. M., Irani, Z., & Weerakkody, V. (2017). Critical analysis of Big Data challenges and analytical methods. *Journal of Business Research, 70*, 263–286. https://doi.org/10.1016/j.jbusres.2016.08.001

Sun, Y., Liu, Y., Zhang, J. Z., Fu, J., Hu, F., Xiang, Y., & Sun, Q. (2021). Dark side of enterprise social media usage: A literature review from the conflict-based perspective. *International Journal of Information Management, 102393*. https://doi.org/10.1016/j.ijinfomgt.2021.102393

Sutherland, W., & Jarrahi, M. H. (2018). The sharing economy and digital platforms: A review and research agenda. *International Journal of Information Management, 43*, 328–341. https://doi.org/10.1016/j.ijinfomgt.2018.07.004

Tawa, J., Ma, R., & Katsumoto, S. (2016). "All lives matter": The cost of colorblind racial attitudes in diverse social networks. *Race and Social Problems, 8*(2), 196–208. https://doi.org/10.1007/s12552-016-9171-z

Tilson, D., Lyytinen, K., & Sørensen, C. (2010). Digital infrastructures: The missing IS research agenda. *Information Systems Research, 21*(4), 748–759. https://doi.org/10.1287/isre.1100.0318

Tretter, F., Simon, K.-H., Reichel, C., Schmaus, T., Serbser, W. H., Hofmann, K. M., & Bieling, C. (2019). Digitalization and sustainability—Human-ecological Perspectives. *GAIA—Ecological Perspectives for Science and Society, 28*(4), 398–400. https://doi.org/10.14512/gaia.28.4.17

Truby, J. (2020). Governing Artificial Intelligence to benefit the UN Sustainable Development Goals. *Sustainable Development, 28*(4), 946–959. https://doi.org/10.1002/sd.2048

United Nations. (2015). *Transforming our World: The 2030 Agenda for Sustainable Development A/RES/70/1.*

Vadisala, J., & Vatsavayi, V. K. (2017). Challenges in social network data privacy. *International Journal of Computational Intelligence Research, 13*(5), 965–979.

van Dam, J.-W., & van de Velden, M. (2015). Online profiling and clustering of Facebook users. *Decision Support Systems, 70*, 60–72. https://doi.org/10.1016/j.dss.2014.12.001

Velásquez, N., Leahy, R., Restrepo, N. J., Lupu, Y., Sear, R., Gabriel, N., Jha, O. K., Goldberg, B., & Johnson, N. F. (2021). Online hate network spreads malicious COVID-19 content outside the control of individual social media platforms. *Scientific Reports, 11*(1), 11549. https://doi.org/10.1038/s41598-021-89467-y

Vinuesa, R., Azizpour, H., Leite, I., Balaam, M., Dignum, V., Domisch, S., Felländer, A., Langhans, S. D., Tegmark, M., & Fuso Nerini, F. (2020). The role of artificial intelligence in achieving the Sustainable Development Goals. *Nature Communications, 11*(1), 233. https://doi.org/10.1038/s41467-019-14108-y

Voicu-Dorobanțu, R., Volintiru, C., Popescu, M.-F., Nerău, V., & Ștefan, G., (2021). Tackling complexity of the just transition in the EU: Evidence from Romania. *Energies, 14*(5), 1509. https://doi.org/10.3390/en14051509

Vraga, E. K., Anderson, A. A., Kotcher, J. E., & Maibach, E. W. (2015). Issue-specific engagement: How facebook contributes to opinion leadership and efficacy on energy and climate issues. *Journal of Information Technology & Politics, 12*(2), 200–218. https://doi.org/10.1080/19331681.2015.1034910

Vu, H. T., Do, H. V., Seo, H., & Liu, Y. (2020). Who leads the conversation on climate change?: A study of a global network of NGOs on Twitter. *Environmental Communication, 14*(4), 450–464. https://doi.org/10.1080/17524032.2019.1687099

Waters, R. D., Burnett, E., Lamm, A., & Lucas, J. (2009). Engaging stakeholders through social networking: How nonprofit organizations are using Facebook. *Public Relations Review, 35*(2), 102–106. https://doi.org/10.1016/j.pubrev.2009.01.006

Weerakkody, V., Sivarajah, U., Mahroof, K., Maruyama, T., & Lu, S. (2021). Influencing subjective well-being for business and sustainable development using big data and predictive regression analysis. *Journal of Business Research, 131*, 520–538. https://doi.org/10.1016/j.jbusres.2020.07.038

Wiese, M., & Akareem, H. S. (2020). Determining perceptions, attitudes and behaviour towards social network site advertising in a three-country context. *Journal of Marketing Management, 36*(5–6), 420–455. https://doi.org/10.1080/0267257X.2020.1751242

World Economic Forum. (2019). *A new circular vision for electronics: Time for a global reboot.* http://www3.weforum.org/docs/WEF_A_New_Circular_Vision_for_Electronics.pdf

Young, W., Russell, S. V., Robinson, C. A., & Barkemeyer, R. (2017). Can social media be a tool for reducing consumers' food waste? A behaviour change experiment by a UK retailer. *Resources, Conservation and Recycling, 117*, 195–203. https://doi.org/10.1016/j.resconrec.2016.10.016

Yu, R. P., Ellison, N. B., & Lampe, C. (2018). Facebook use and its role in shaping access to social benefits among older adults. *Journal of Broadcasting & Electronic Media, 62*(1), 71–90. https://doi.org/10.1080/08838151.2017.1402905

Yu, R. P., Mccammon, R. J., Ellison, N. B., & Langa, K. M. (2016). The relationships that matter: Social network site use and social wellbeing among older adults in the United States of America. *Ageing & Society, 36*(9), 1826–1852. https://doi.org/10.1017/S0144686X15000677

Zutshi, A., Grilo, A., & Nodehi, T. (2021). The value proposition of blockchain technologies and its impact on Digital Platforms. *Computers & Industrial Engineering, 155*, 107187. https://doi.org/10.1016/j.cie.2021.107187

The Influence of Social Participation for Sustainable Against Regression: A Critical Review of Brazilian Environmental Public Policies in Light of the Environmental Justice

Paulo Santos de Almeida◉ and Vitor Calandrini de Araujo◉

Abstract This work tries to contribute to the discussion of the importance of the influence of Social Participation toward the regression in the relationship between sustainability and Environmental Justice through the Brazilian Environmental Public Policies, both in the field of research and in environmental management. The points in common between the concepts of sustainability and Environmental Justice through the Brazilian Environmental Public Policies are analyzed, as well as the critical factors that hinder its practical application against the real regression without inclusion the Social Participation in a democratic society, as a starting point to move toward a synergistic strategy on the local power and community of the people in status vulnerabilities model for sustainable life, into the called Smart Sustainability. This proposal is based on building a true and transparent governance and is aimed at taking advantage of the opportunities offered by the public policies for a more efficient and inclusive sustainable management.

Keywords Social Participation · Sustainable against Regression: Brazilian Environmental Public Policies · Environmental Justice · Inclusion

1 Introduction

This work focuses on the influence of environmental public policy in Brazil's actions that make the building of social participation and Environmental Justice. Specifically, it studies the influence of public environmental actions. This paper investigates and criticizes the impact of the phenomenon of social participation is increasingly involved with the strongest inherent in the co-creation and participation process.

Social networks facilitate individual participation and extend their role within organizations. These developments have impacted relationships between citizens and stakeholders. Encouraging this kind of collaboration has become an even bigger

P. S. de Almeida (✉) · V. C. de Araujo
Programa de Pós-graduação em Sustentabilidade, Escola de Artes, Ciências e Humanidades, Universidade de São Paulo, Rua Arlindo Bettio, 1000, São Paulo, Brasil
e-mail: psalmeida@usp.br

challenge for managers in extremely competitive environments such as in democratic systems.

On another hand, this overview aims the incidence derived from having no synergies between participation and environmental regression on the public policies, resulting in hazardous to Human Rights, well-being and economic.

The essay's organization is as follows. First Section is devoted to the literature review in order to identify the research gap and view about Sustainability and Participation. Second part presents the relationship and the results from Brazilian Environmental Law and SDG. The discussion and conclusion are in the last sections. The results indicate that social participations have several dependency actions and building the link between the public policies, sustainability, participation and strategies of governance influenced by socio-environmental.

2   Social Participation as the Essential Element Toward Sustainability

Important to be aware of the difficulties that exist today to make feasible, for example, proposals that articulate the reduction of environmental degradation with income generation. Although this theme is the object of projects guided by the local political will of administrators, the intentionality is not always successful or well or well understood by the residents, especially in vulnerability status. These are systems that require a period of maturation and whose legitimization is quite slow, on the part of the various social levels, but they need to have been built with participating actions previewed by law.

Successful decisions and experiences, mainly on the part of administrations and the Judiciary, if acting directly in the interests of vulnerable local communities, they show that if there is political will it has been possible to make feasible governmental actions guided by the adoption of the principles of environmental sustainability that prevent retrogression and result in effects in the sphere of economic, social and environmental development (Jacobi, 1999). They also show that it is in the municipality that it is easier to develop the combined action of several integration programs with social participation. This makes it possible to articulate policies of regional character with extension of the good quality of life on common goods and fundamental rights, such as access to health and environmental balance.

2.1   Sustainability and Its Ways

Sustainable development sometimes has many issues and it needs a kind of figure out and sensibility of knowledge into this concept. Nowadays Sustainability could make confusion on Sustainable development because it is inherent to a group of environmental researchers. It seems the communities have a feeling that sustainability and sustainable development are the same substance, but beyond the great conceptual problem they can consider more involvement of people and governments to resolve all the real problems to consumptions and use of environmental research and commons.

After the 1980s, the concept was more useful because it was implemented in Brundtland Report, Our Common Future. The great building was made in 1992 in Rio de Janeiro and performed in the UN Conferences Environment and sustainable development through its principles toward the sustainability debate on the global fields input the current Agenda.[1]

Sustainable development is about the desire to grant every person in the world to lead a decent life, combating poverty and hunger, freedom, democracy, safety, and Human Rights; there should be a chance for each human being to be a full member of society. As well as the concerns, and consciousness that there are limits of the over exploring on our planet and people (Roorda, 2016).

The progress is being made in many places, but, overall, action to meet the Goals is not yet advancing at the speed or scale required. 2020 needs to usher in a decade of ambitious action to deliver the Goals by 2030.

Allying this demand came up the "Decade of Action," which calls for accelerating sustainable solutions to all the world's biggest challenges. All sectors of society mobilize for a decade of action on three levels: global action to secure greater leadership, more resources and smarter solutions for the Sustainable Development Goals; local action embedding the needed transitions in the policies, budgets, institutions and regulatory frameworks of governments, cities and local authorities; and people action, including by youth, civil society, the media, the private sector, unions, academia and other stakeholders, to generate an unstoppable movement pushing for the required transformations (UN, 2021).

The *COVID-19* pandemic and its impact on all 17 SDGs have reduced risks from future potential crises and relaunched the implementation efforts to deliver the 2030 Agenda and SDGs during the current "Decade of Action - 2020 -2030." The Decade of Action will mobilize everyone, everywhere because these demand urgency and ambition to supercharge ideas to solutions in the world.[2]

2.2 The Inadmissible Regression on Environmental Sustainability

The society cannot address trends without homogenizing all guarantees, reducing differences by imposing an integrative process in order to against injustices and inequalities, although building and enforcement it an equity that transmits, while maintaining thus capacity, the document model social interest in common goods in harmony balanced with the environment (Flores, 2010).

---

[1] https://www.un.org/en/conferences/environment/rio1992.

[2] https://www.un.org/sustainabledevelopment/decade-of-action/.

The protection of minorities must exist as conditions for political participation so that pragmatists do not distance social interests, prevent the full exercise of the State's constitutional guarantees as well as the guarantee of inclusion and reduction of vulnerability, promoting social participation in all its instances whether considering democracy as an essential tool for this purpose.

Never the regress of governmental visions should impede integration, inclusion, or the Human Rights that were hard-won in modern society. There is a growing obligation of commitment through modern States to guarantee the rights and dignity of the human person.

Regression prohibition is a general principle of Environmental Law. As pointed out Benjamin (2011), in Brazil it is safe to say that the prohibition of retrogression, despite not being, with name and surname, enshrined in the Federal Constitution, nor in infra-constitutional norms, and notwithstanding its relative imprecision—understandable in recently formulated institutes and still in full process of consolidation—became a general principle of Environmental Law, to be invoked in the assessment of the legitimacy of legislative initiatives aimed at reducing the level of legal protection of the environment, especially in what affects in particular (a) essential ecological processes, (b) fragile ecosystems or on the brink of collapse and (c) species threatened with extinction as pointed out by Benjamin (2011).

Prieur (2011) describes it as a risk of "non-regression" because the terminology used by the doctrine is still hesitant. In certain countries, there is talk of a standstill principle. In France, the concept of cliquet (lock) effect is used, or the anti-re-crease cliquet rule tour (non-return lock). Non-retrogression is similarly assimilated to the theory of acquired rights, when the latter can be attacked by regression. It also evokes the "irreversibility," note—in the field of Human Rights. These provisions, like all other fundamental rights, are furthermore regulated by Articles 53 and 54 of the EU Charter of Fundamental Rights. The Charter cannot be interpreted as "limiting" recognized rights, nor as implying the right to destroy or limit them beyond what is foreseen. In this case, the provisions reinforce the obligation of non-regression and, thus, the prohibition of retrogression in the legal protection of the environment. These are classic clauses in Human Rights conventions, such as Articles 17 and 53 of the European Convention on Human Rights.

3   How to Improve the Social Participation in Public Policies in Brazil?

The construction of social participation in Brazil has some alternatives in the Brazilian legal system. The Federal Constitution of 1988 (Brazil, 1988) has as an essential characteristic in the social participation since its creation in 1986. The so-called "citizen's charter" was characterized by democratization with the breaking of the dictatorial period between 1964 and 1984 to direct the social reaction with the guarantee of individual and collective rights, and art. 225 guarantees environmental protection and conservation as a common right of everyone.

The Public Hearings are recurrent in the systematic information of administrative procedures, as in the Environmental Impact Assessment and Ecological-Economic Zoning foreseen by the National Environmental Policy - Act n° 6.938/81.

In this hand, the best way to support a strategy toward effective social participation within governance transparency beyond the governability or such of governmental actions, there are efficient conditions their that set up whether governance could be integrated inherently with good practice and partnership with the society (participation), that show us the results are probable to accept them and affordable for the people.

> Yet governance—or misgovernance—may be less than transparent, or near to entirely repressed. While effective governance also may be rent by conflictual institutional logics. Such as, in public health services, conflicts between market-driven managerial concern to reduce costs and professional concern to ensure high quality clinical care. (Oliveira et al., 2016, p. 163)

3.1 Building the "Localizing" of Agenda 2030

"Localizing" is the process of taking into account subnational contexts in the achievement of the 2030 Agenda, from the setting of goals and targets, to determining the means of implementation and using indicators to measure and monitor progress. Localization relates both to how the SDGs can provide a framework for local development policy and to how local and regional governments can support the achievement of the SDGs through action from the bottom up and to how the SDGs can provide a framework for local development policy (Global Taskforce of Local & Regional Governments, 2016).

As a global strategy toward promoting equity in the world, the SDGs have 17 goals to push on the socio-environment at the top of the Earth Agenda from 2015 to next decade. Is it possible? How could the UN make this difference?

There are no easy answers that could be done through simple or unique planning, these plans must be architectured always with the support of the society (private sector, governments, ONG). Although this is a herculean task, it is not impossible. In this scenario the subnational public policy is the key.

The Public Administrations has the administrative machine to improve, and also push it on the society through a large and transparent governance based on efforts focused on the in-depth social participation. But it need to be on the direct qualitative relationship between social subsystems and their influencing factors in local interests for the sustainable development, driving to suitability to provide a powerful integrated practical matter's solutions as mobility, water's access, or waste's ecological destination. All of it has been demanded in towns and communities.

3.2 The Social Participation of the Vulnerabilities and Environmental Justice

The degree of socioeconomic vulnerability is associated with differential exposure to risks and indicates greater or lesser exposure of people, places, infrastructure and/or ecosystems to some particular type of harm, configuring an unequal distribution of risks not only socially, but spatially (Canil et al., 2021).

There are these Principles of Environmental Justice (Environmental Working Group n/d):

1. Environmental justice affirms the sacredness of Mother Earth, ecological unity and the interdependence of all species, and the right to be free from ecological destruction.
2. Environmental justice demands that public policy be based on mutual respect and justice for all peoples, free from any form of discrimination or bias.
3. Environmental justice mandates the right to ethical, balanced and responsible uses of land and renewable resources in the interest of a sustainable planet for humans and other living things.
4. Environmental justice calls for universal protection from nuclear testing, extraction, production and disposal of toxic/hazardous wastes and poisons and nuclear testing that threaten the fundamental right to clean air, land, water, and food.
5. Environmental justice affirms the fundamental right to political, economic, cultural and environmental self-determination of all peoples.
6. Environmental justice demands the cessation of the production of all toxins, hazardous wastes, and radioactive materials, and that all past and current producers be held strictly accountable to the people for detoxification and the containment at the point of production.
7. Environmental justice demands the right to participate as equal partners at every level of decision-making including needs assessment, planning, implementation, enforcement and evaluation.
8. Environmental justice affirms the right of all workers to a safe and healthy work environment, without being forced to choose between an unsafe livelihood and unemployment. It also affirms the right of those who work at home to be free from environmental hazards.
9. Environmental justice protects the right of victims of environmental injustice to receive full compensation and reparations for damages as well as quality health care.
10. Environmental justice considers governmental acts of environmental injustice a violation of international law, the Universal Declaration On Human Rights, and the United Nations Convention on Genocide.
11. Environmental justice must recognize a special legal and natural relationship of Native Peoples to the U.S. government through treaties, agreements, compacts, and covenants affirming sovereignty and self-determination.

12. Environmental justice affirms the need for urban and rural ecological policies to clean up and rebuild our cities and rural areas in balance with nature, honoring the cultural integrity of all our communities, and providing fair access for all to the full range of resources.
13. Environmental justice calls for the strict enforcement of principles of informed consent, and a halt to the testing of experimental reproductive and medical procedures and vaccinations on people of color.
14. Environmental justice opposes the destructive operations of multi-national corporations.
15. Environmental justice opposes military occupation, repression and exploitation of lands, peoples and cultures, and other life forms.
16. Environmental justice calls for the education of present and future generations which emphasizes social and environmental issues, based on our experience and an appreciation of our diverse cultural perspectives.
17. Environmental justice requires that we, as individuals, make personal and consumer choices to consume as little of Mother Earth's resources and to produce as little waste as possible; and make the conscious decision to challenge and reprioritize our lifestyles to insure the health of the natural world for present and future generations. Adopted today, October 27, 1991, in Washington, D.C.

The discussion of Environmental Justice becomes inevitable and necessary (Acserald, 2002); an approach that raises the issue of the need to advance in public policies, mainly to address the essential aspects to increase and confirm the resilience in social interaction with the tendency of increases in disastrous and severe events that will affect the most vulnerable groups (Canil et al., 2021; Lampis et al., 2020). This requires, for the most part, those most vulnerable from the affected population to react and resist in managing socio-environmental risks. There is, therefore, the construction of an intellectual process of the organized society of NGO, and the Public Administration in the conduct and application of public policies, both based on dialogue and interaction to improve the practices of social participation in a cooperative manner (Sulaiman, 2018).

Table 1 shows that as vulnerability increases, there is more backlash and less public awareness. On the other hand, it is observed that the more participation there

Table 1 Relationship of scale in local public power, participation, regression and its results commons, and public awareness

Scale/Actions	Local	Regression	Results	Awareness
Vulnerability	+	+	−	−
Participation	+	−	+	+
Envtl. Justice	−	+	−	−

Source The authors

is, the less regression occurs. Finally, it is noted that the lack of local participation leads to a lack of environmental justice. This situation is undoubtedly the worst scenario because it is the absence of perspective, guarantees and a better quality of life, especially for vulnerable people.

4 The Debate of Inclusion and Sustainability Against the Regression on Environmental Public Policies

Currently, the sustainability debate is confronted with the retrocession of environmental public policy systems because it restricts the vitality of democracy. Since 2020, in Brazil, it has been observed that inefficiency has compromised balanced governance in sectors that require greater transparency and effectiveness, such as the economic, social, health and environmental sectors.

The sustainability requires a World alignment that allows a harmonization in multilateral diplomacy that contributes to the evolutionary process of protection and conservation of these sectors, but that in particular can positively change the quality of life of people in situations of greater vulnerability, protecting them and allowing them to progress and prosperity in a balanced way through minimum and fundamental rights and guarantees. These measures require greater political and social integration in order to maintain the necessary perspectives and adjustments.

The Brazilian public policies have reached a level of effectiveness that directed Brazilian society toward higher levels of the democratic quality of life for people, including respecting non-human life with greater protection.

However, only with the democratic improvement of public policies combined with the efforts of civil society will institutions and organizations be strengthened to adapt their capacities, opportunities and well-being to sustainable and inclusive development, reshaping future solutions to situations of poverty reduction and vulnerability, especially for the most affected and systematically discriminated communities.

4.1 The Real Discussion in Brazil of the Sustainability After COVID-19 Pandemic Period (2020/2021)

Although there are compatible rights protected constitutionally, sustainability and sustainable development walk in the search for balance, effectiveness and guarantee of the fundamental rights of quality of life, and human dignity with economic equity. They are, therefore, fundamental rights protected in the Brazilian Federal Constitution (Brazil, 1988).

As pointed out by Leff (2003) who discussed the environmental complexity and the evolution of society that emerged with another rationality and another way of thinking about production and consumption, looking at the forecast made in Stockholm (UN, 1972) because the ethical conditions are fundamental to understand and also question the cultural and social development of each era.

> Environmental knowledge is the questioning of the ecological conditions of sustainability and the social bases of democracy and justice; it is a construction and communication of knowledge that puts on display the judgment of power strategies and the effects of domination that are generated through the holding, appropriation, and transmission of knowledge. (Leff, 2003)

In this sense, the *Covid-19* pandemic has led us to think about the Environmental approach in an even more integrated way, differentiating our reality as weak sustainability or strong sustainability, always associated with the rescue of the importance of the limits of environmental resources as opposed to maintaining thinking based exclusively on the neoliberal economic system.

The diversities have been exposed in the economic-social, cultural and environmental senses. Besides the misfortune of hundreds of thousands of deaths, humanity will still have to adapt to a new reality and a new way of understanding sustainability. The Public Police will be obliged to build and articulate its potentialities with the organized society in order to effect the reduction of inequalities, and with this, the protection of the most vulnerable subjects.

Thus, the consequences of the *Covid-19* pandemic revealed the lack of concept of the sustainability because "while it is possible to have goods from all over the world delivered quickly and cheaply to one's doorstep, given the heavy concentration of supply chains in a few low-wage countries like China and just-in-time production, the global economic system" also lacks any rescue system to absorb any disruptions, as has now happened rather violently in the current pandemic. In other words, it is not a very "resilient" system as point out Machado and Richter (2020), and it has to be, although.

4.2 The Strategy of Inclusion Against the Regression of the People on the Vulnerabilities' Status

Poverty is a condition historically determined by social provisions. There is a concern in quantitatively measuring poverty without understanding the phenomenon of its generation and reproduction, thus, its definition cannot be based only on statistical data relating to income and purchasing control but must also consider the political and social dimension to which it is linked (Santos, 2009). It will be necessary to have data that make the vulnerability of individuals compatible so that it is not seen as a condition of economic deprivation but as a way of life that combines social, economic, cultural and political (Fracalanza et al., 2013; Santos, 2009) and environmental relations as well.

The differences are aggravated when public resources are not directed to the implementation of inclusive social policies. With globalization and the growing need for labor for the maintenance of flows of products and services, keeping migratory movements continuously between less economically favored regions fostering the migration of the low-income population in peripheral areas of cities, devoid of minimum infrastructures, such as access to basic sanitation or health services. With this, the dynamics of the metropolis keep the vulnerable away from the best opportunities for access to quality of life. In this sense, a significant portion of socio-environmental problems is magnified by the lack of development policy suitable for a real and concrete social equity, thus neglecting the rights and driving injustice (Fracalanza et al., 2013).

5 Conclusions

This study analyzed and criticized the sustainability, the social participation and its influence on Brazilian Environmental Public Policies, and also sustainable development of its aspects, including the demonstration that the Environmental Justice without social participation, sustainability has no effective development.

The ways of sustainability just could be completed through the inclusion of the society by formal and also informal social participation. Starting off the legal system into and for the Brazilian Federal Constitution, which previously had a large and structured guarantee of rights in a mandatory architecture that it has been to promote citizen participation. The inclusion of society is perhaps the most interesting facet of dignity and citizenship, because the dialogues have relied upon one of the essential constitution goals since it was promulgated in 1988 (Brazil, 1988).

After and with the SDG Agenda as a global strategy to make the difference between past and present for the currently and the future generations. As localizing considers the process of taking into account subnational contexts in the achievement of the 2030 Agenda, from the setting of goals and its targets, the socio-environmental gained a kind of protagonism toward push on the scenario, and have to make strongly to keep it on the public policies.

The Brazilian Federal Constitution guarantees that public management must promote a more inclusive and democratic environmental policy, regardless of the governments that deconstruct the gains obtained against the regressions.

It was observed that the relationships between *COVID-19* pandemic and sustainability to people status' vulnerabilities in their view addressed more impact than others with the direct effect that it causes loss and poverty. This negative intersection has been seen by disrespect in listening in each community.

In the post-pandemic era, society must be more positive and participative in building an affordable tomorrow to share the commons with vulnerabilities above the personal interests.

References

Acselrad, H. (2002). Justiça ambiental e construção social do risco. *Desenvolvimento e meio ambiente, 5,* 49–60.

Canil, Katia et al. (2021). *Vulnerabilidades, Riscos E Justiça Ambiental em Escala Macro Metropolitana. Mercator (Fortaleza)* [online]. 20 e20003. Epub 15 March 2021. ISSN 1984–2201. https://www.scielo.br/j/mercator/a/zbBrtD9Fx963k7WCf8TwLRy/#. Accessed 12 November 2021.

Benjamin, A. H. (2011). *Princípio da Proibição de Retrocesso Ambiental* (Brasil. Congresso Nacional. Senado Federal. Comissão de Meio Ambiente, Defesa do Consumidor e Fiscalização e Controle - CMA). O princípio da proibição de retrocesso ambiental. Colóquio Internacional sobre o Princípio da Proibição de Retrocesso Ambiental (2012: Brasília, DF) (p. 55). Brasília: Senado Federal. http://www2.senado.leg.br/bdsf/handle/id/242559. Accessed 8 October 2021.

Brazil. (1988). *Federal Constitution of 05 October 1988.* http://www.planalto.gov.br/ccivil_03/constituicao/constituicao.htm. Accessed 30 October 2021.

Brazil. Congresso Nacional. Senado Federal. Comissão de Meio Ambiente, Defesa do Consumidor e Fiscalização e Controle - CMA. (2011). *O princípio da proibição de retrocesso ambiental.* Colóquio Internacional sobre o Princípio da Proibição de Retrocesso Ambiental (2012: Brasília, DF) (p. 269). Brasília: Senado Federal. http://www2.senado.leg.br/bdsf/handle/id/242559. Accessed 8 October 2021.

de bem Machado, A., & Richter, M. F. (2020). Sustentabilidade em Tempos de Pandemia (Covid-19). *Recima21 - Revista Científica Multidisciplinar, 1*(2), 264–279. ISSN 2675–6218. https://doi.org/10.47820/recima21.v1i2.25. Accessed 12 November 2021.

Environmental Working Group—EWG. *The principle of environmental justice.* Online. https://www.ewg.org/news-insights/news/17-principles-environmental-justice. Accessed 17 November 2021.

Flores, J. H. (2010). La Construcción De Las Garantías. Hacia Una Concepción Antipatriarcal De La Libertad Y La Igualdad In D. Sarmento, D. Ikawa, & F. Piovesan (Eds.), *Igualdade, Diferença e Direitos Humanos.* Lumen Juris.

Fracalanza, A. P., Jacob, A. M., & Eça, R. F. (2013, January–March). Justiça Ambiental e Práticas de Governança da Água: (Re) Introduzindo Questões De Igualdade Na Agenda. *Ambiente & Sociedade São Paulo, XVI*(1), 19–38. https://www.scielo.br/j/asoc/a/Yzc9zvrxWYCb4LyFWm4r4yq/?lang=pt&format=pdf. Accessed 10 November 2021.

Global Taskforce of Local and Regional Governments. (2016). *Roadmap for localizing the SDGs: Implementation and monitoring at subnational level.* https://sustainabledevelopment.un.org/content/documents/commitments/818_11195_commitment_ROADMAP%20LOCALIZING%20SDGS.pdf. Accessed 11 March 2021.

Jacobi, P. (1999). Poder local, políticas sociais e sustentabilidade. *Saúde e Sociedade, 8*(1). online published 05 June 2008. https://www.scielo.br/j/sausoc/a/db4rjM8KWWZgP5TttCTXfXk/?format=html&lang=pt. Accessed 1 November 2021.

Lampis, A, Campello, P. T., Jacobi, P. R., Leonel, A. L. (2020). A produção de riscos e desastres na América Latina em um contexto de emergência climática. *O Social em Questão, 48,* 75–92. https://www.researchgate.net/publication/343794626_A_producao_de_riscos_e_desastres_na_America_Latina_em_um_contexto_de_emergencia_climatica. Accessed 12 November 2021.

Leff, E. (2003). Pensar a Complexidade Ambiental. In E. Leff (Coord.), *A complexidade ambiental.* Cortez.

Oliveira, T. C., Raposo, V., Holland, S., & de Carvalho, F. E. L. (2016). From new public management to new public services: Challenges for hospital governance and lean and hybrid management. In C. Machado & J. P. Davim (Eds.), *Green and lean management* (p. 163). Springer.

Prieur, M. (2011). *O Princípio da Proibição de Retrocesso Ambiental* (Brasil. Congresso Nacional. Senado Federal. Comissão de Meio Ambiente, Defesa do Consumidor e Fiscalização e Controle - CMA). O princípio da proibição de retrocesso ambiental. Colóquio Internacional sobre o Princípio da Proibição de Retrocesso Ambiental (2012: Brasília, DF) (p.11). Senado Federal. http://www2.senado.leg.br/bdsf/handle/id/242559. Accessed 8 October 2021.

Roorda, N. (2016). The seven competences of a sustainable professional: The RESFIA+D model for human resource management (HRm), educations and training. In C. Machado & J. P. Davim (Eds.), *Management for sustainable development*. Publisher River. ISBN 9788793379084.

Santos, M. (2009). *Pobreza Urbana*. Edusp.

Sulaiman, S. N. (2018). Reflexão e ação: educar para uma cultura preventiva. In S. N. Sulaiman & P. R. Jacobi (Eds.), *Melhor prevenir: olhares e saberes para a redução de risco de desastre* (p. 23). IEE-USP. http://www.incline.iag.usp.br/data/arquivos_download/melhorprevenir_ebook.pdf. Accessed 12 November 2021.

United Nations. (1972, June 5–16). *United Nations conference on the human environment*. Stockholm. https://undocs.org/en/A/CONF.48/14/Rev.1. Accessed 11 November 2021.

United Nations. (2021). *Decade of action: Sustainable development goals* (online). https://www.un.org/sustainabledevelopment/decade-of-action/. Accessed 10 November 2021.

Analysis of Food Loss and Waste for the European Countries in the Context of Sustainable Development

Adrian Stancu⦿

Abstract Nowadays, in the context of green economy, wastes represent the concern of most of the world countries. Considering also the food loss and food waste, which means discarded food, and given that, at this moment, there are people suffering from hunger, this issue has become of major importance. At the beginning of the chapter, a new approach is set for the delimitation of food loss and food waste concepts. The reduction of the food loss and waste is a goal of numerous sustainable development strategies. Thus, particular measurement systems and databases of food loss and waste employed by the Eurostat, the United States Environmental Protection Agency, and the Food and Agriculture Organization of the United Nations are described. The analysis focuses on the animal and mixed food waste, vegetal wastes, and household and similar wastes generated by households and economic activities in the European countries between 2010 and 2018. The results underscore the clusters of countries with similar behavior of food loss and waste and the link between the country's economic development and the quantity of food waste generated by households and economic activities.

Keywords Food loss · Food waste · Sustainable development · European countries · Economic development

1 Introduction

The humanity has been striving against hunger since ancient Egypt (McDermott, 2001), ancient Greece (Garnsey, 2004), etc. and will be striving against hunger at least until 2030 according to the United Nations' Sustainable Development Goals (United Nations, 2021).

In 2020, on average, 768 million people around the world faced hunger with 118 million people more than in 2019 due to the COVID-19 pandemic (FAO et al., 2021). The countries from Africa South of the Sahara and those from South Asia recorded

A. Stancu (✉)
Petroleum-Gas University of Ploiesti, Ploiesti, Romania
e-mail: astancu@upg-ploiesti.ro

the highest level of Global Hunger Index (GHI) score in 2020. In countries such as Chad, Timor-Leste, Madagascar, Burundi, Central African Republic, Comoros, the Democratic Republic of the Congo, Somalia, South Sudan, Syria, and Yemen, the hunger is considered alarming (von Grebmer et al., 2020). Meanwhile, 45% of fruits and vegetables, 45% of roots and tubers, 35% of fish and seafood, 30% of cereals, 20% of dairy products, 20% of meat, and 20% of oilseeds and pulses are lost or wasted worldwide every year (Food & Agriculture Organization of the United Nations, 2015). These are average values because the food losses depend on various factors, such as for the grains: their type, their variety, the harvest method, the time of harvesting, the weather conditions, meeting or not the criteria for food, etc. (Hartikainen et al., 2017). With respect to the above-mentioned antinomic figures, measuring in depth and diminishing the food loss and waste represent a major priority.

Since 2011, the Food and Agriculture Organization of the United Nations (FAO) has underscored the difference between food loss and food waste based on the stage in the food supply chain where the wastes occur (Food & Agriculture Organization of the United Nations, 2011). Thus, both food loss and food waste consist in reductions in the quantity or quality of food resulting from decisions and actions taken by the food suppliers in the case of food loss, and by retailers, food services, and consumers for food waste (Food and Agriculture Organization of the United Nations, 2019a) (Fig. 1a).

The same outlook was shared by the World Bank (2020), the World Economic Forum (2018) and also by some researchers (Alabrese et al., 2015; Corvellec, 2013;

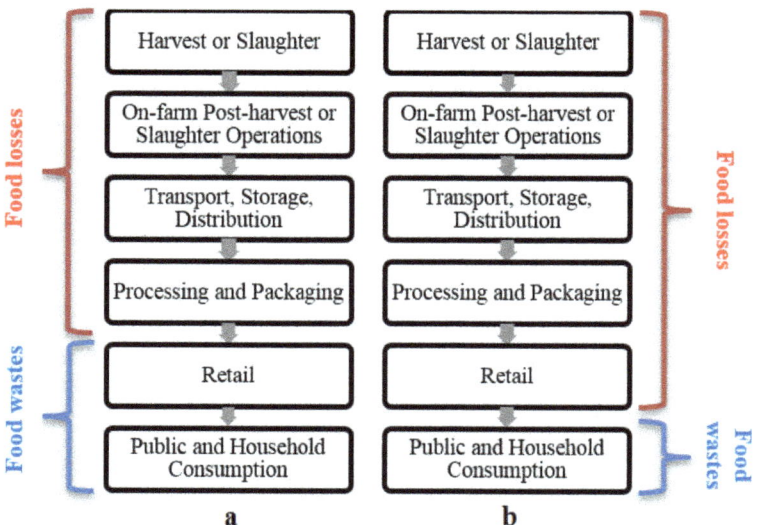

Fig. 1 Differences between food loss and food waste from FAO's perspective (a) and author's perspective (b) (*Source* Author's development based on Food and Agriculture Organization of the United Nations [2019b])

Kummu et al., 2012; Schneider, 2013). However, the food waste generated only by households is part of the post-consumer waste (Ene, 2013).

I think that FAO's approach is not accurate because the wastes that arise from all economic activities of the food supply chain must be included in only one category, i.e., "food loss". Hence, the "food waste" category comprises the waste generated only by households, hospitals, colleges, universities, office buildings, military bases, etc., known as general "consumers" (Fig. 1b).

In a previous report, FAO asserted that food waste (which is linked with deliberate discarding) may arise at all stages of the food supply chain (Food & Agriculture Organization of the United Nations, 2017). This statement was introduced by Parfitt et al. (2010) and it is not appropriate because all the waste ranging from harvesting or slaughtering to the public and household consumption must be included in the food loss category regardless of the type of discarding. Otherwise, the grouping of food loss and food waste by the stage of the supply chain has no sense.

Similarly, from the standpoint of the food supply chain, the food loss and waste are indicators of the Food Chain Inefficiency (FCI) which emphasize the state of the process in which they arise. In addition to this, food loss is related to the unavoidable cases, independent from the human will, and the food waste is linked to the avoided issues in the supply chain (Horton et al., 2019; Pinstrup-Andersen et al., 2016).

Food loss and waste are connected only with the products for human consumption and are not applicable in the case of non-edible feeds and parts of products. If an edible product becomes non-edible at any stage of the supply chain, it is framed as food loss or waste (Food & Agriculture Organization of the United Nations, 2011), which is a suitable approach.

Human over-consumption, viewed as over-nutrition, i.e., the gap between a higher food availability in contrast to a low food need, is considered food waste (Horton et al., 2019; Smil, 2004).

The High Level Panel of Experts on Food Security and Nutrition (2014) discussed the new concept of food quality loss and waste (FQLW) which is the reduction of the food quality at any stage of the supply chain, from harvest or slaughter to consumption. Lagioia et al. (2021) brought in the notion of food wastage which is caused by both food loss and food waste.

The fact that food loss and waste are part of the sustainable development is confirmed by the 2030 Agenda for Sustainable Development (2030 ASD) which was adopted in 2015 by the member states of the United Nations. The 2030 ASD comprises 17 Sustainable Development Goals (SDGs), 169 targets, and 230 indicators. The 12th SDG (Ensure sustainable consumption and production patterns) and more precisely the Target 12.3 (By 2030, halve per capita global food waste at the retail and consumer levels and reduce food losses along production and supply chains, including post-harvest losses) emphasize the importance of food loss and waste for a sustainable development (United Nations, 2015, 2021). Thus, food waste management is essential for ensuring food safety and improving public health (Blakeney, 2019; Ene et al., 2017).

The aims of the chapters are, firstly, to describe the measurement system and the database of food loss and waste used in Europe, in the United States of America, and by specific international organizations in order to identify the corresponding available data. Secondly, to analyze the food waste in the European countries between 2010 and 2018 to emphasize the level of each country as compared to the others. Thirdly, to highlight the cluster of countries with similar behavior of food loss and waste and to underline the interrelatedness of the country's economic development and the quantity of food waste generated by households and economic activities.

2   Food Loss and Waste Measurement

The importance of food loss and waste in the context of sustainable development generated the need for measurement by regional and international institutions. The measurement methodology used by the European Union (EU), the United States Environmental Protection Agency (EPA), and FAO will be presented in the following.

2.1   Measurement by Eurostat

In the EU, the Regulation No 2150 of the European Parliament and of the Council, also known as Waste Statistics Regulation (WStatR), stated that Eurostat should manage statistics of data waste by categories, including food waste (European Commission, 2002). Thus, the food waste database was established and first data was reported by both EU and non-EU countries starting with 2004 for one waste category and beginning with 2010 for the other two waste categories. Furthermore, there were adopted complementary legal acts such as Directive 2008/98/EC (Waste Framework Directive) which established the reporting obligation (European Commission, 2008) and the Decision 2011/753/EU which stipulated the rules and calculation methods (European Commission, 2011).

In 2012, Eurostat created the "food waste plug-in" initiative to increase the accuracy in collecting food waste data. Thus, this measure resulted into the following WStatR main waste categories used to measure the food waste (Eurostat, 2021a):

- Animal and mixed food waste;
- Vegetal wastes;
- Household and similar wastes.

The European Waste Classification for Statistics (EWC-Stat) divided the above-mentioned waste categories into some List of Waste (LoW) sub-categories. Thus, the animal and mixed food waste category comprises 10 waste sub-categories (such as, the animal-tissue waste, the materials unsuitable for consumption or processing, the biodegradable kitchen and canteen waste, etc.), the vegetal wastes category includes 11 waste sub-categories (for instance, the wastes from forestry, the sludge from washing and cleaning, the plant-tissue waste, etc.), and household and similar wastes category contains 5 waste sub-categories (e.g., the mixed municipal waste, the waste from markets, the street-cleaning residues, etc.) (Eurostat, 2013, 2021a).

Each of these three waste types can be generated by numerous sectorial activities and thus grouped into 8 main categories considering the statistical classification of economic activities in the European Community (NACE Rev. 2) by Eurostat (2021b) (Fig. 2).

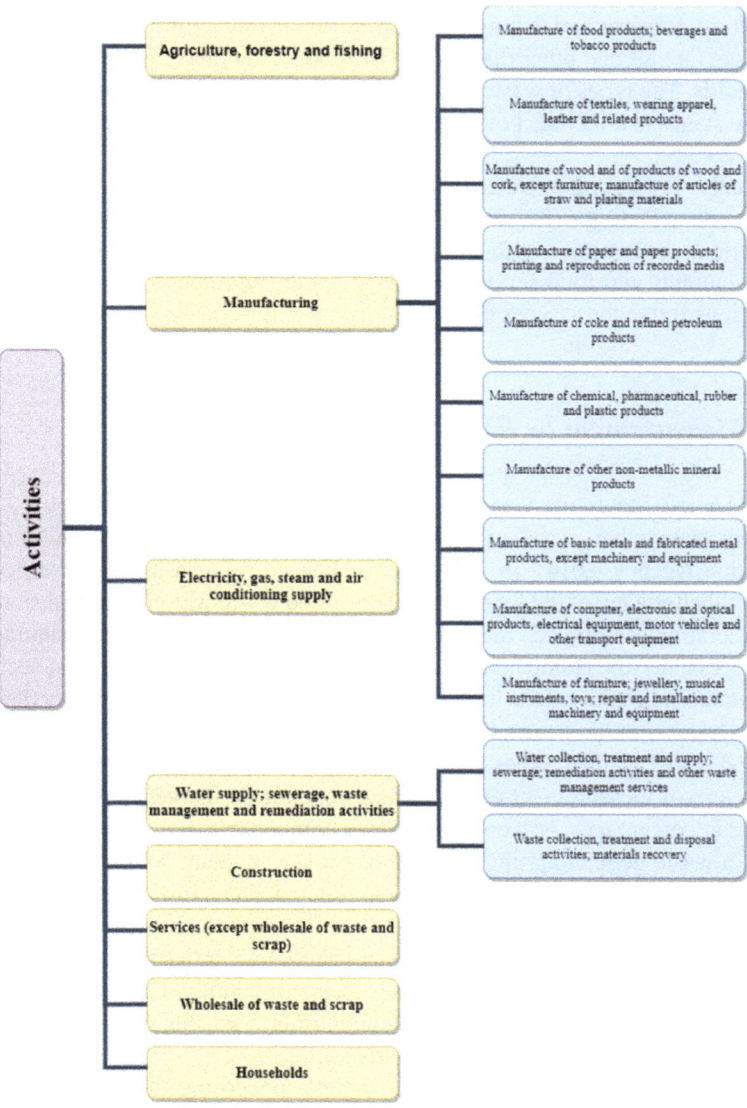

Fig. 2 Activities that generate food waste by NACE Rev. 2 (*Source* Author's development based on Eurostat [2021b])

A brief analysis of the Waste Framework Directive, Eurostat's food waste database and NACE Rev. 2 highlights that, in the EU, there is not a clear delimitation between food loss and food waste data reported by the countries.

Starting with 2019, the common methodology and minimum quality requirements for the uniform measurement of levels of food waste and the format for reporting of data on food waste were adopted as an improvement of the existing methodological framework. Along with this new approach, the food waste is individually measured in the following five stages of the food supply chain: primary production (which include agriculture, forestry, and fishing), processing and manufacturing (for both foods and beverages), retail and other distribution of food, restaurants and food services, and households. The types of waste which include food waste are also specified for each of these stages and include both animal and vegetal wastes (European Commission, 2019a, 2019b).

Equally Commission Delegated Decision (EU) 2019/1597 and Commission Implementing Decision (EU) 2019/2000 still define the waste generated from primary production, processing and manufacturing, retail and other distribution of food, and restaurants and food services as "food waste" and not as "food loss" as other institutions frame these types of waste.

2.2   Measurement by EPA

In the United States of America, EPA has measured food waste, as a separate category from other waste categories, since 1960 (United States Environmental Protection Agency, 2014). In 2017, EPA enhanced its measurement methodology to have a better view of the food waste. Thus, food waste was measured according to five generating sectors, i.e., the industrial sector (food and beverage manufacturing and processing), the residential sector (family housing), the commercial sector (the food retail/wholesale and the hospitality sectors), the institutional sectors (hospitals, nursing homes, military installations, office buildings, colleges, universities, etc.), and food banks (United States Environmental Protection Agency, 2020a).

The unit of measurement of food waste depends on the generating sector and it is either in percentages, or it comprises three components, i.e., a unit of mass (lbs or tones), a waste generating unit (person, employee, student, household, etc.), and a unit of time (year), e.g., lbs/household/year (United States Environmental Protection Agency, 2020b).

The limitation of the EPA measurement methodology is that there is no breakdown of the food waste category to allow the distinction among different types of food waste, i.e., vegetal waste, animal waste, etc. Furthermore, the methodology does not measure the food loss from the agricultural sector.

So far, EPA has published only one report concerning the food waste which contains data from 2018. Hence, a comparative analysis of the values from 2018 with the previous years cannot be done.

2.3   Measurement by FAO

FAO manages a food loss and waste database (FLW Database) with some important options in contrast to Eurostat. Firstly, data collected starts with 1945 and it ends with 2017 (Food and Agriculture Organization of the United Nations, 2021a). Even if data is spread for more than 70 years, there is only one record in 1945 (from India regarding the losses in storage of wheat), and the following year with data available is 1961 for 22 countries. Secondly, data is calculated in percentages, and it relates to all world countries which can be individually selected or examined in aggregate form, e.g., there are 10 country groups, such as SDG Region, Least Developed Countries, World Bank Income Groups, etc. (Food and Agriculture Organization of the United Nations, 2021b).

Thirdly, the database contains loss and waste for 146 different types of foods which are grouped in five categories, as follows (Food and Agriculture Organization of the United Nations, 2021a):

- Cereals and pulses;
- Fruits and vegetables;
- Roots, tubers, and oil-bearing crops;
- Meat and animal products;
- Other.

Albeit there is no distinct option in the FLW Database interface to choose between food loss and food waste, the selection can be done by picking the value chain stage where the loss or waste occurred. There are 17 standard value chain stages, i.e., consumer, distribution, export, grading, harvest, market, packaging, pre-harvest, processing, producer, restaurants, retail, stacking, storage, traders, transport, and wholesale, and additional two options, namely, parameter estimate and total supply chain estimate (Food and Agriculture Organization of the United Nations, 2021a).

Fourthly, the FLW Database comprises data collected from 18 methods, for instance, case studies, census, survey, expert opinion, FAO sources, food balance sheet, national statistics yearbook, filed test and trail, etc. (Food and Agriculture Organization of the United Nations, 2021a). This option may represent an advantage if the aim is to obtain data from a unique source. However, it turns into an important disadvantage if all sources are considered, due to the data heterogeneity.

Although the FLW Database offers numerous selection options, it comprises discontinuous time series. For instance, by applying the filters to identify the waste for all 146 types of foods generated by consumers from all world countries between 2010 and 2017, collected by all methods, the output is a file with only 13 records, i.e., 8 records from the United States of America, 3 records from Finland, and one record from Norway and Kazakhstan. The year of food waste data is between 2010 and 2012 in the case of the United States of America, between 2010 and 2014 for Finland, 2014 for Kazakhstan, and 2017 for Norway (Food and Agriculture Organization of the United Nations, 2021a). Considering this state of fact, data provided by FLW Database cannot be analyzed.

Based on the Target 12.3 from 2030 ASD, FAO developed the indicator 12.3.1 (Global Food Loss and Waste) which consist of two sub-indicators, namely, 12.3.1.a—Food Loss Index and 12.3.1.b—Food Waste Index (Food and Agriculture Organization of the United Nations, 2021c).

The Global Food Loss Index (GFLI) and the Regional Food Loss Index (RFLI) are computed on the aggregation of the country-level Food Loss Indices (FLI), by commodity, on annual frequency. GFLI, RFLI, and FLI are measuring the food loss from the production to the retail stage, namely the supply side of the value chain (Food and Agriculture Organization of the United Nations, 2021d, 2021e; Gennari, 2015).

Food Waste Index is covering the food retail, households, and food services sectors, that is the demand side of the value chain. It is calculated distinctly for each of three sectors, by country, at a global level (United Nations Environment Programme, 2021).

In the latest FAO's report on Food Waste Index (United Nations Environment Programme, 2021) and other documents related to GFLI, RFLI, and FLI (Fabi, 2020, Food and Agriculture Organization of the United Nations, 2019c, 2021d, 2021f) there are no data which can be analyzed.

3 Analysis of Food Waste in the European Countries

In this section the food waste for 38 European countries is analyzed between 2010 and 2018 based on data reported every two years by each state. Thus, the term "previous year" used in this chapter means two years behind the year that is analyzed. The time-series data used was collected from Eurostat (2021b) and is limited to year 2018, even if the last table structure change was in February 2021 and the last update of data was in June 2021 (Eurostat, 2021c).

The analysis was conducted for each of the three waste types (animal and mixed food waste, vegetal wastes, and household and similar wastes) but considering one single category of hazardous and non-hazardous waste. The activities that generate waste were grouped by author into two main categories based on the NACE Rev. 2 (Fig. 2), i.e., *households* and *economic activities*. The second category was set up by the author to bound data reported by the countries and to establish the differences between food loss and food waste, and it comprises the following activities:

- Agriculture, forestry, and fishing;
- Mining and quarrying;
- Manufacturing;
- Electricity, gas, steam, and air conditioning supply;
- Water supply; sewerage, waste management, and remediation activities;
- Construction;
- Services (except wholesale of waste and scrap);
- Wholesale of waste and scrap

To ensure the intelligibility of the results, the graph of each food waste category that was analyzed excludes the countries that recorded null values or have not reported any data. But this case was mentioned for each state.

3.1   Analysis of Animal and Mixed Food Waste

3.1.1   Analysis of Animal and Mixed Food Waste Generated by Households

Concerning the animal and mixed food waste generated by households (which can be framed as food waste) in the European countries, in 2010, the highest values were for the Netherlands (76 kg per capita), Italy (42 kg per capita), Luxembourg (39 kg per capita), Norway (35 kg per capita), and Austria (20 kg per capita). By opposite, the lowest values were in the case of Denmark and Belgium (1 kg per capita), Estonia (4 kg per capita), Slovenia and the United Kingdom (5 kg per capita), Iceland (7 kg per capita), and Ireland (11 kg per capita), except the countries that recorded null values or have not reported any data (Albania, Bosnia and Herzegovina, Bulgaria, Croatia, Cyprus, Czechia, France, Germany, Greece, Hungary, Kosovo, Latvia, Liechtenstein, Lithuania, Malta, Montenegro, North Macedonia, Poland, Portugal, Romania, Serbia, Slovakia, and Turkey). In 2012, the countries with the highest levels were the Netherlands (78 kg per capita), Italy (52 kg per capita), Luxembourg (47 kg per capita), Austria (43 kg per capita), and Norway (35 kg per capita), while the states with the lowest level were Belgium (1 kg per capita), Denmark (3 kg per capita), Estonia (4 kg per capita), Iceland and the United Kingdom (8 kg per capita), and Slovenia (10 kg per capita). The countries that recorded null values or have not reported any data in 2010 were the same in 2012 (Fig. 3).

In 2014, the Netherlands (79 kg per capita), Italy (61 kg per capita), Luxembourg (49 kg per capita), Austria (43 kg per capita), and Finland (41 kg per capita) recorded the highest values, in contrast to Belgium, Bulgaria, Iceland, Serbia, and Slovakia (1 kg per capita), Denmark and Estonia (4 kg per capita), Montenegro and the United Kingdom (7 kg per capita), Ireland and Slovenia Slovakia (10 kg per capita), and Spain (12 kg per capita) that registered the lowest values. The states that recorded null values or have not reported any data in 2012 were the same in 2014 except for Bulgaria, Montenegro, Serbia, and Slovakia.

In 2016, the top 5 countries with the highest and lowest values were slightly different as compared to 2014, namely, the Netherlands (82 kg per capita), Italy (74 kg per capita), Finland (48 kg per capita), Austria (44 kg per capita), and Norway (36 kg per capita) for the first category, and Belgium, Croatia, Czechia, Hungary, Iceland, and Serbia (1 kg per capita), Bulgaria (2 kg per capita), Estonia (4 kg per capita), Malta and Slovakia (5 kg per capita), and Denmark and the United Kingdom (9 kg per capita) for the second category. The countries that recorded null values or have not reported any data in 2014 were the same in 2016 except for Croatia, Czechia, Hungary, and Malta. Finally, in 2018, the Netherlands (86 kg per capita),

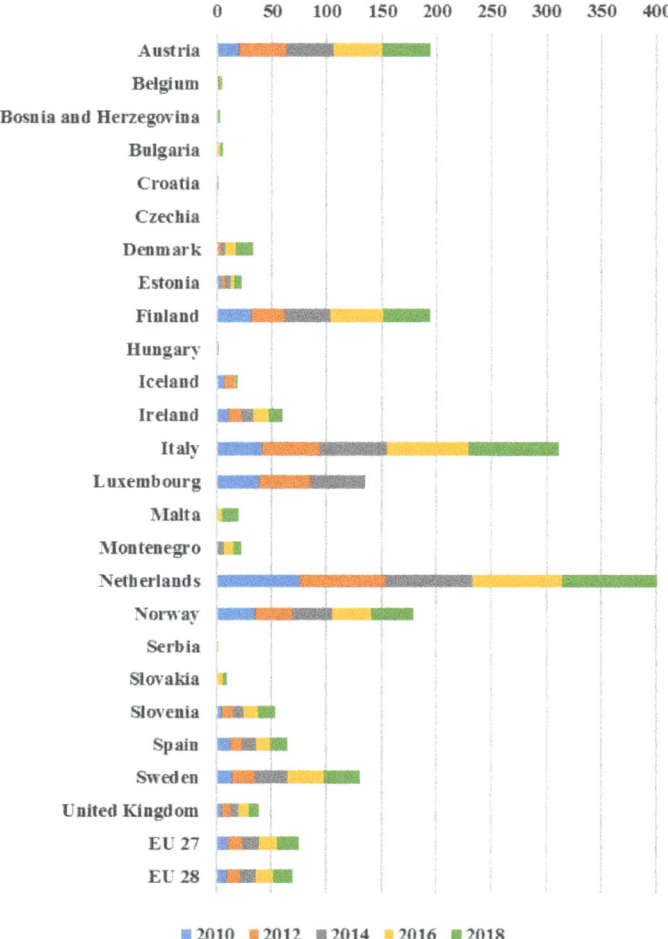

Fig. 3 Animal and mixed food waste generated by households in the European countries between 2010 and 2018 (kilograms per capita) (*Source* Author's development based on Eurostat [2021b])

Italy (83 kg per capita), Austria (44 kg per capita), Finland (43 kg per capita), and Norway (38 kg per capita) recorded the highest level. On the contrary, Belgium, Croatia, and Hungary (1 kg per capita), Iceland (2 kg per capita), Bulgaria, Bosnia and Herzegovina (3 kg per capita), Slovakia (4 kg per capita), and Estonia (7 kg per capita) registered the lowest level. Czechia, Luxembourg, and Serbia recorded null values in contrast to 2016, whereas Bosnia and Herzegovina reported the data in 2018 for the first time.

The analysis of the evolution of the animal and mixed food waste generated by households in the European countries, between 2010 and 2018, underscores a different country ranking. Thus, in 2012 in contrast to 2010, the highest increases were in the case of Denmark (+200%), Austria (+115%), Slovenia (+100%), the

United Kingdom (+60%), and Sweden (+50%) and, by contrast, the lowest increases were for the Netherlands (+2.63%), Ireland (+9.09%), Iceland (+14.29%), Luxembourg (+20.51%), and Italy (+23.81%). There were no states that recorded decreases and Albania, Belgium, Bosnia and Herzegovina, Bulgaria, Croatia, Cyprus, Czechia, Estonia, Finland, France, Germany, Greece, Hungary, Kosovo, Latvia, Liechtenstein, Lithuania, Malta, Montenegro, North Macedonia, Norway, Poland, Portugal, Romania, Serbia, Slovakia, Spain, and Turkey maintained a constant level of their waste. In 2014 against 2012, there were only six increases, i.e., the highest rises were for Sweden (+42.86%), Denmark (+33.33%), and Finland (+32.26%), and the lowest increases were for the Netherlands (+1.28%), Luxembourg (+4.26%), and Italy (+17.31%). It is the first year when the quantity of the animal and mixed food waste generated by households declined, in the case of Iceland (−87.50%), Ireland (−16.67%), and the United Kingdom (−12.5%). Albania, Austria, Belgium, Bosnia and Herzegovina, Croatia, Cyprus, Czechia, Estonia, France, Germany, Greece, Hungary, Kosovo, Latvia, Liechtenstein, Lithuania, Malta, North Macedonia, Norway, Poland, Portugal, Romania, Slovenia, Spain, and Turkey recorded the same level.

In 2016 as compared to 2014, the highest growths were achieved by Slovakia (+400%), Denmark (+125%), Bulgaria (+100%), Ireland (+40%), and Slovenia (+30%), and the lowest increases were recorded by Austria (+2.33%), Norway (+2.86%), the Netherlands (+3.8%), Spain (+8.33%), and Sweden (+10%). In particular, Luxembourg was the only state with a reduction of food waste (−100%), and Albania, Belgium, Bosnia and Herzegovina, Cyprus, Estonia, France, Germany, Greece, Iceland, Kosovo, Latvia, Liechtenstein, Lithuania, North Macedonia, Poland, Portugal, Romania, Serbia, and Turkey registered a constant level of waste. In 2018 against 2016, Malta (+200%), Iceland (+100%), Denmark (+77.78%), Estonia (+75%), and Bulgaria (+50%) registered the highest increase and the Netherlands (+4.88%), Norway (+5.56%), Italy (+12.16%), the United Kingdom (+11.11%), and Slovenia and Spain (+23.08%) recorded the lowest rise. Czechia (+100%), Serbia (−100%), Slovakia (+20%), Finland (−10.42%), and Ireland (−7.14%) registered declines. Albania, Austria, Belgium, Croatia, Cyprus, Hungary, Kosovo, Latvia, Liechtenstein, Lithuania, Luxembourg, Montenegro, North Macedonia, Poland, Portugal, Romania, Sweden, and Turkey generated a similar level of waste.

The analysis highlights that 2 (Austria and Finland) out of 38 countries recorded both the highest level of waste and significant increase as compared to previous year, 5 (Austria, Italy, Luxembourg, the Netherlands, and Norway) out of 38 countries recorded both the highest level of waste and irrelevant rise as compared to previous year, 2 (Finland and Luxembourg) out of 38 countries recorded both the highest level of waste and decline against previous year, 9 (Bulgaria, Denmark, Estonia, Iceland, Ireland, Malta, Slovakia, Slovenia, and the United Kingdom) out of 38 countries recorded both the lowest level of waste and a major increase versus previous year, 5 (Iceland, Ireland, Slovenia, Spain, and the United Kingdom) out of 38 countries recorded both the lowest level of waste and insignificant rise as opposed to previous year, and 5 (Czechia, Iceland, Ireland, Serbia, and Slovakia) out of 38 countries recorded both the lowest level of waste and decrease as compared to the previous year.

3.1.2 Analysis of Animal and Mixed Food Waste Generated by Economic Activities

As regards the animal and mixed food waste generated by economic activities (which can be enclosed as food loss), firstly, in 2010, the highest levels were for Liechtenstein (533 kg per capita), Ireland (354 kg per capita), Iceland (307 kg per capita), Norway (125 kg per capita), and Belgium (108 kg per capita). North Macedonia and Turkey (1 kg per capita), Serbia (3 kg per capita), Bulgaria (4 kg per capita), Romania (6 kg per capita), and Czechia (7 kg per capita) recorded the lowest level. Albania, Bosnia and Herzegovina, Kosovo, and Montenegro recorded null values or have not reported any data. In 2012, the highest level was reached by Iceland (380 kg per capita), Ireland (193 kg per capita), Norway (116 kg per capita), Belgium and Finland (103 kg per capita), and France (63 kg per capita). On the contrary, the following cases can be highlighted with the lowest level: Turkey (1 kg per capita), Serbia (3 kg per capita), Bulgaria (4 kg per capita), Romania (6 kg per capita), and Czechia (7 kg per capita). Albania and North Macedonia recorded null values or have not reported any data (Fig. 4).

In 2014, Belgium (96 kg per capita), Ireland (73 kg per capita), Finland (70 kg per capita), Liechtenstein (65 kg per capita), and France (61 kg per capita) were the countries with the highest level. Conversely, North Macedonia (1 kg per capita), Serbia and Turkey (2 kg per capita), Italy (5 kg per capita), Czechia, Portugal, Romania (7 kg per capita), and Croatia and Hungary (8 kg per capita) recorded the lowest level. Albania and Kosovo recorded null values or have not reported any data. In 2016, the highest levels were for Ireland (183 kg per capita), Belgium (105 kg per capita), Luxembourg (83 kg per capita), Liechtenstein (77 kg per capita), and Finland (73 kg per capita), while, by contrast, the lowest levels were for Turkey (1 kg per capita), North Macedonia and Serbia (2 kg per capita), Montenegro (4 kg per capita), Italy (6 kg per capita), and Czechia (7 kg per capita).

In 2018, Ireland (187 kg per capita), Belgium (108 kg per capita), Liechtenstein (97 kg per capita), Iceland (83 kg per capita), and Luxembourg (78 kg per capita) recorded the highest levels, and Turkey (2 kg per capita), Bosnia and Herzegovina (3 kg per capita), Serbia (4 kg per capita), Bulgaria and North Macedonia (6 kg per capita), and Italy (7 kg per capita) reported the lowest levels. In both 2016 and 2018, Albania has not reported any data.

Secondly, the evolution of the animal and mixed food waste generated by economic activities emphasized that, in 2012 against 2010, Bulgaria (+275%), France (+90.91%), Finland (+33.77%), Iceland (+23.78%), Sweden (+22%), and Lithuania (+19.35%) recorded the highest rises, and Slovakia (+11.11%), Slovenia (+18.52%), and Malta (+12.5%) registered the lowest increases. The highest declines were in the case of North Macedonia (−100%), Estonia (−92.86%), Liechtenstein

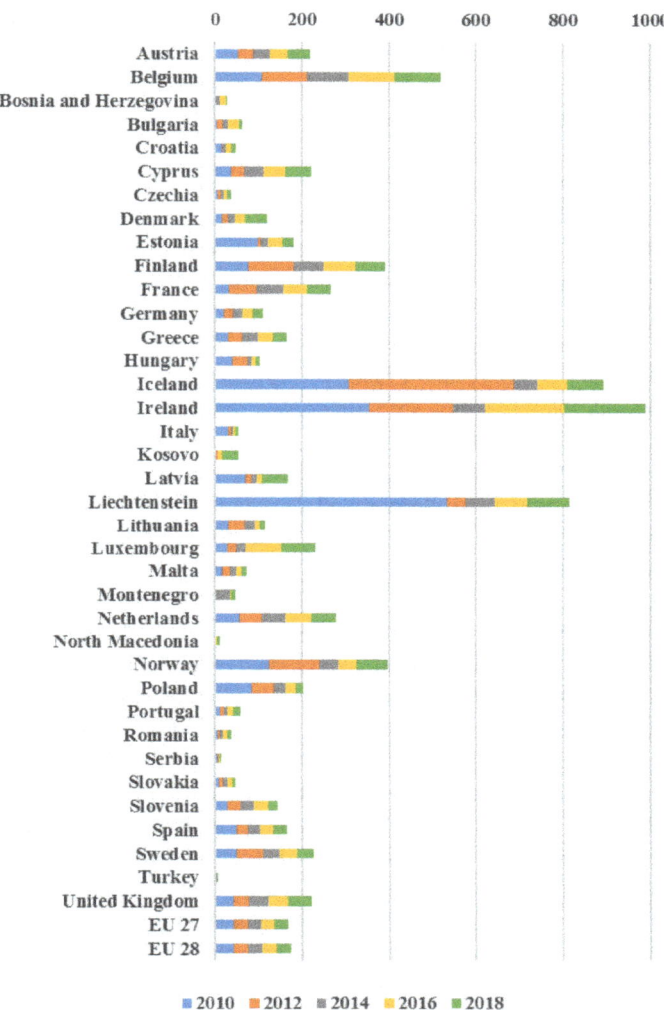

Fig. 4 Animal and mixed food waste generated by economic activities in the European countries between 2010 and 2018 (kilograms per capita) (*Source* Author's development based on Eurostat [2021b])

(−91.93%), Italy (−83.87%), and Croatia (−80%), and the lowest falls were for the Netherlands (−1.82%), Denmark (−6.25%), Norway (−7.2%), Portugal (−9.09%), and Hungary (−10%). Albania, Czechia, Germany, Greece, Montenegro, Romania, Serbia, and Turkey maintained a constant level of their waste. In 2014 versus 2012, the following cases with the highest increases can be highlighted: Bosnia and Herzegovina (+400%), Croatia (+166.67%), Estonia (+157.14%), Turkey (+100%), and Liechtenstein (+54.16%). The lowest growths were for Sweden (+3.85%), Germany (+4.76%), Luxembourg (+5%), Austria (+5.71%), and Denmark (+6.67%). On the contrary, Kosovo (−100%), Iceland (−86.05%), Hungary (−77.78%), Norway (−63.79%), and Ireland (−62.18%) recorded the highest diminutions, whereas France (−3.17%), Slovenia (−6.25%), Belgium (−6.8%), Slovakia (−10%), and Malta (−11.11%) registered the lowest decreases. Albania, Czechia, Italy, Latvia, and the Netherlands reported the same level.

In 2016 as compared to 2014, the highest increases were for Luxembourg (+295.24%), Ireland (+150.68%), Bulgaria (+145.45%), North Macedonia (+100%), and Portugal (+85.71%), and the lowest boosts were for the United Kingdom (+2.27%), Sweden (+2.63%), Finland (+4.29%), Germany (+4.55%), and Slovenia (+6.67%). Montenegro (−88.24%), Turkey (−50%). Lithuania (−47.83%), Malta (−31.25%), and Latvia (−23.08%) recorded the highest reductions, and Greece (−5.41%), Poland (−7.69%), and France (−11.48%) registered the lowest declines. Albania, Czechia, Norway, and Serbia recorded the same level of waste. In 2018 as opposed to 2016, the following countries with the highest rises can be underlined: Latvia (+490%), Kosovo (+227.27%), North Macedonia (+200%), Denmark (+117.39%), and Montenegro, Serbia, and Turkey (+100%). The lowest increases were in the case of Ireland (+2.19%), Belgium (+2.86%), France (+3.7%), Lithuania (+8.33%), and Malta (+9.09%). Bulgaria (−77.78%), Bosnia and Herzegovina (−76.92%), Slovakia (−41.67%), Poland (−33.33%), and Slovenia (−28.13%) recorded the highest cuts, and Finland (−5.48%), Luxembourg (6.02%), Spain (−6.25%), the Netherlands (−6.78%), and Greece (−11.43%) registered the lowest falls. Albania, Croatia, Germany, and Sweden supplied a similar level of waste.

The analysis underlined the following cases: 6 (Finland, France, Iceland, Ireland, Liechtenstein, and Luxembourg) out of 38 countries registered both the highest level of waste and important increase as compared to previous year, 4 (Belgium, Finland, France, and Ireland) out of 38 countries recorded both the highest level of waste and minor rise as compared to previous year, 4 (Iceland, Ireland, Liechtenstein, and Norway) out of 38 countries recorded both the highest level of waste and significant decline as against previous year, 8 (Bosnia and Herzegovina, Bulgaria, Croatia, Montenegro, North Macedonia, Portugal, Serbia, and Turkey) out of 38 countries recorded both the lowest level of waste and a significant increase versus previous year, and 9 (Bosnia and Herzegovina, Bulgaria, Croatia, Hungary, Italy, Kosovo, Montenegro, North Macedonia, and Turkey) out of 38 countries recorded both the lowest level of waste and a major decrease as opposed to previous year. There were no countries which recorded both the lowest level of waste and slight increase as opposed to previous year.

3.2   Analysis of Vegetal Wastes

3.2.1   Analysis of Vegetal Waste Generated by Households

Relating to the vegetal waste generated by households (which can be framed as food waste) in the European countries, in 2010, Belgium (109 kg per capita), Germany

(104 kg per capita), Luxembourg (95 kg per capita), Denmark (89 kg per capita), and the United Kingdom (63 kg per capita) recorded the highest values. Conversely, Romania (1 kg per capita), Poland (2 kg per capita), Estonia and Latvia (3 kg per capita), Hungary (10 kg per capita), and Malta (11 kg per capita) registered the lowest values. Countries such as Albania, Bosnia and Herzegovina, Bulgaria, Croatia, Cyprus, Kosovo, Finland, Greece, Liechtenstein, Lithuania, Montenegro, North Macedonia, Portugal, Serbia, Spain, and Turkey recorded null values or have not reported any data. In 2012, the top 5 states with the highest levels were Germany (113 kg per capita), Denmark (105 kg per capita), Belgium (98 kg per capita), Luxembourg (95 kg per capita), and Austria (62 kg per capita), while the top 5 countries with the lowest level were Lithuania, Romania, and Turkey (1 kg per capita), Poland (3 kg per capita), Croatia and Cyprus (8 kg per capita), Malta (11 kg per capita), and Ireland (16 kg per capita). At the same time, Albania, Bosnia and Herzegovina, Bulgaria, Kosovo, Finland, Greece, Hungary, Liechtenstein, Montenegro, North Macedonia, Portugal, Serbia, and Spain recorded null values or have not reported any data in 2012 (Fig. 5).

In 2014, Liechtenstein (197 kg per capita), Germany (119 kg per capita), Denmark (112 kg per capita), Belgium (85 kg per capita), and Luxembourg (80 kg per capita) recorded the highest values, as opposed to Romania (1 kg per capita), Bulgaria, Estonia, and Lithuania (4 kg per capita), Latvia (8 kg per capita), Croatia (9 kg per capita), and Serbia Latvia (11 kg per capita) which registered the lowest values. The null values and no reported data were in the following cases: Albania, Bosnia and Herzegovina, Kosovo, Finland, Greece, Hungary, Portugal, Spain, and Turkey. In 2016, the highest values were for Liechtenstein (178 kg per capita), Germany (121 kg per capita), Denmark (110 kg per capita), Austria (72 kg per capita), and Belgium (70 kg per capita), and the lowest values were for Romania (2 kg per capita), Estonia (4 kg per capita), Croatia, Malta, and Serbia (5 kg per capita), Luxembourg (8 kg per capita), and Ireland (15 kg per capita). Albania, Bosnia and Herzegovina, Kosovo, Finland, Hungary, Latvia, Portugal, Spain, and Turkey recorded null values or have not reported any data.

In 2018, Liechtenstein (165 kg per capita), Germany and Denmark (118 kg per capita), Austria (74 kg per capita), Belgium (67 kg per capita), and France and the United Kingdom (61 kg per capita) recorded the highest values. By contrast, Serbia (1 kg per capita), Romania (4 kg per capita), Bosnia and Herzegovina, Latvia, and Malta (5 kg per capita), Croatia, Greece, and Spain (6 kg per capita), and Bulgaria and Estonia (9 kg per capita) registered the lowest values. The null values and no reported data were in the case of Albania, Kosovo, Hungary, and Turkey.

Secondly, the evolution of the vegetal waste generated by households highlighted that, in 2012 against 2010, Latvia (+1,000%), Estonia (+166.67%), Austria (+77.14%), Slovenia (+60%), and Poland (+50%) recorded the highest increases, while Norway (+3.03%), the Netherlands (+3.7%), Slovakia (+5.88%), the United Kingdom (+6.35%), and Germany (+8.65%) registered the lowest growths. The only decreases were in the case of Hungary (−100%), Luxembourg (−18.95%), Austria (−10.09%), and Sweden (−3.03%). Albania, Bosnia and Herzegovina, Bulgaria, Finland, Greece, Ireland, Italy, Kosovo, Liechtenstein, Malta, Montenegro, North

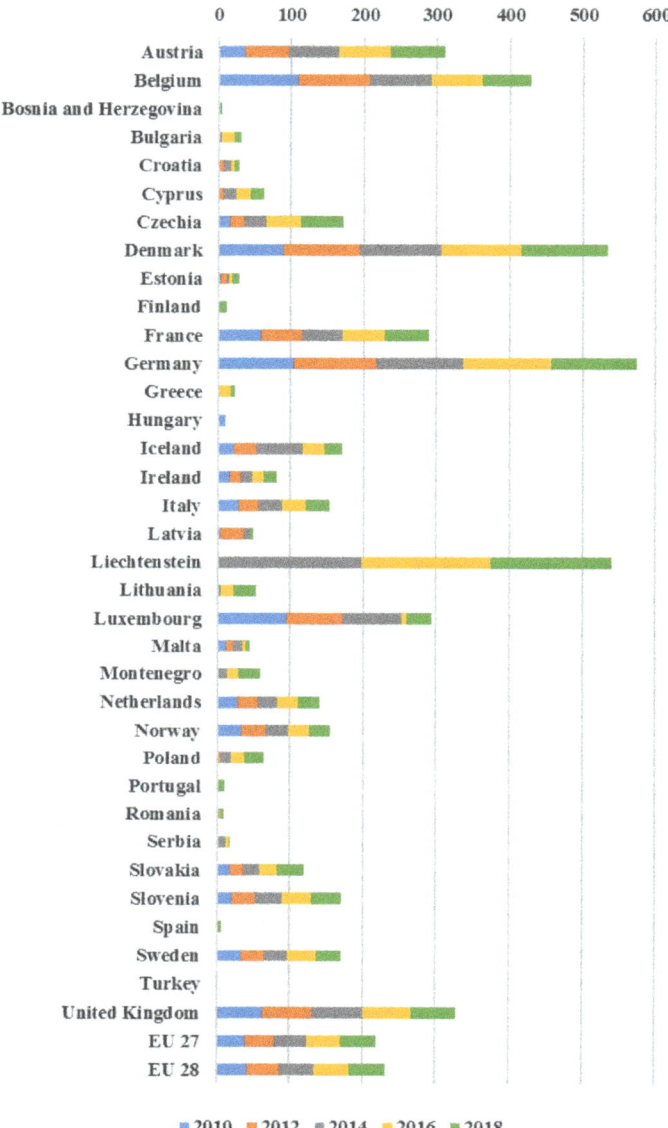

Fig. 5 Vegetal waste generated by households in the European countries between 2010 and 2018 (kilograms per capita) (*Source* Author's development based on Eurostat [2021b])

Macedonia, Portugal, Romania, Serbia, and Spain maintained a constant level of their waste. In 2014 as compared to 2012, highest increases were for Poland (+333.33%), Lithuania (+300%), Cyprus (+112.5%), Iceland (+106.45%), and Czechia (+50%), and the lowest rises were for Luxembourg (+3.9%), the United Kingdom (+4.48%), Germany (+5.31%), Denmark (+6.67%), and Austria (+8.06%). There were only seven declines in the case of Turkey (−100%), Latvia (−75.76%), Estonia (−50%), Belgium (−13.27%), Norway (−8.82%), Ireland (−6.25%), and France (−1.75%). Albania, Bosnia and Herzegovina, Finland, Greece, Hungary, Kosovo, the Netherlands, North Macedonia, Portugal, Romania, Spain, and Sweden have not changed their level of waste.

In 2016 as opposed to 2014, the following cases with the highest growths can be underscored: Bulgaria (+350%), Lithuania (+300%), Romania (+100%), Czechia (+56.67%), and Poland (+46.15%). Germany (+1.68%), France (+1.79%), Italy (+3.13%), Slovakia (+4.35%), and Austria (+7.46%) recorded the lowest increases. The highest falls were for Latvia (−100%), Luxembourg (−90%), Malta (−58.33%), Iceland (−54.69%), and Serbia (−54.55%). There were no differences among the levels from 2016 versus 2014 in the case of Albania, Bosnia and Herzegovina, Estonia, Finland, Hungary, Kosovo, Ireland, the Netherlands, North Macedonia, Portugal, Spain, and Turkey. In 2018 in contrast to 2016, Luxembourg (+312%), Estonia (+125%), Lithuania and Romania (+100%), Montenegro (+81.25%), and Slovakia (+54.17%) recorded the highest expansions and Slovenia (+2.5%), Austria (+2.78%), the Netherlands (+3.57%), France (+7.02%), and Denmark (+7.27%) registered the lowest increases. Conversely, Serbia (−80%), Greece (−64.71%), Bulgaria (−50%), Iceland (−13.79%), and Sweden (−12.82%) recorded the highest declines. The following states maintained constant their level: Albania, Cyprus, Hungary, Kosovo, Italy, Malta, North Macedonia, and Turkey.

The analysis underscored that one (Luxembourg) out of 38 countries recorded both the highest level of waste and significant increase as compared to previous year, 6 (Austria, Denmark, France, Germany, Luxembourg, and the United Kingdom) out of 38 countries recorded both the highest level of waste and irrelevant rise as compared to previous year, 7 (Belgium, Denmark, France, Germany, Liechtenstein, Luxembourg, and the United Kingdom) out of 38 countries recorded both the highest level of waste and decline against previous year, 8 (Bulgaria, Cyprus, Estonia, Latvia, Lithuania, Luxembourg, Poland, and Romania) out of 38 countries recorded both the lowest level of waste and a major increase versus previous year, one (Luxembourg) out of 38 countries recorded both the lowest level of waste and insignificant rise as opposed to previous year, and 11 (Bulgaria, Croatia, Estonia, Greece, Hungary, Ireland, Latvia, Luxembourg, Malta, Serbia, and Turkey) out of 38 countries recorded both the lowest level of waste and decrease as compared to previous year.

3.2.2   Analysis of Vegetal Waste Generated by Economic Activities

As for to the vegetal waste generated by economic activities (which can be enclosed as food loss), in 2010, the higher levels were in the case of the Netherlands (523 kg per capita), Belgium (216 kg per capita), Lithuania (95 kg per capita), Denmark (69 kg per capita), and Cyprus (68 kg per capita), and the lowest levels were for Iceland (2 kg per capita), Turkey (5 kg per capita), Latvia and Malta (6 kg per capita), Greece (8 kg per capita), and North Macedonia (9 kg per capita). Albania, Bosnia and Herzegovina, Kosovo, Montenegro, and Norway recorded null values or have not reported any data. In 2012, the Netherlands (517 kg per capita), Belgium (234 kg per capita), Lithuania (118 kg per capita), Austria (77 kg per capita), and Bulgaria (70 kg per capita) reported the highest levels, and Montenegro (1 kg per capita), Iceland and North Macedonia (3 kg per capita), Bosnia and Herzegovina (5 kg per capita), Croatia (6 kg per capita), and Kosovo, Malta, Norway, and Portugal (7 kg per capita) registered the lowest levels (Fig. 6).

In 2014, the following states with the highest level can be underlined: the Netherlands (503 kg per capita), Belgium (320 kg per capita), Lithuania (148 kg per capita), Finland (72 kg per capita), and Austria (67 kg per capita). At the opposite end, Kosovo (1 kg per capita), North Macedonia (2 kg per capita), Liechtenstein (5 kg per capita), Bosnia and Herzegovina, Iceland, and Turkey (6 kg per capita), and Malta and Norway (7 kg per capita) recorded the lowest levels. In 2016, the Netherlands (496 kg per capita), Belgium (407 kg per capita), Kosovo (122 kg per capita), Luxembourg (118 kg per capita), and Lithuania (113 kg per capita) registered the highest levels, and North Macedonia (1 kg per capita), Iceland (3 kg per capita), Turkey (5 kg per capita), Norway (6 kg per capita), and Bosnia and Herzegovina, Portugal, and Serbia (7 kg per capita) recorded the lowest levels.

In 2018, the highest levels were for Belgium (590 kg per capita), the Netherlands (516 kg per capita), Luxembourg (101 kg per capita), Kosovo (100 kg per capita), and Austria (83 kg per capita), and in the case of the lowest levels were Bosnia and Herzegovina (1 kg per capita), Iceland (2 kg per capita), North Macedonia (3 kg per capita), Liechtenstein (4 kg per capita), and Malta, Norway, and Serbia (6 kg per capita). Albania was the only country that has not reported any data between 2012 and 2018.

Secondly, the evolution of the vegetal waste generated by economic activities underlined that, in 2012 compared to 2010, Bulgaria (+125.81%), Hungary and Latvia (+100%), Turkey (+60%), Austria (+50.98%), and Iceland (+50%) recorded the highest rises, and Germany (+9.09%), Belgium (+12.5%), Slovenia (+3.33%), Malta (+16.67%), and Lithuania (+24.21%) registered the lowest growths. On the contrary, the highest diminutions occurred in the case of Estonia (−83.02%), Italy (−80.77%), Cyprus (−76.47%), North Macedonia (−66.67%), and Denmark (−56.52%). In 2014 against 2012, the highest increases were for Montenegro (+1,600%), Croatia (+200%), Latvia (+166.67%), Estonia (+122.22%), and Iceland (+100%), and the lowest rises were for France (+2.08%), Romania (+9.76%), Sweden (+18.52%), Bosnia and Herzegovina (+20%), and Germany (+20.83%). Kosovo (−85.71%), Liechtenstein (−82.14%), Poland (−45.31%), Bulgaria (−41.43%), and North Macedonia (−33.33%) recorded the highest drops.

In 2016 in contrast to 2014, the highest boosts were in the case of Kosovo (+12,100%), Luxembourg (+461%), Greece (+137.5%), Montenegro (+305.88%), and Latvia (+96.88%), and the lowest increases were for Denmark (+2.08%), France (+6.12%), the United Kingdom (+6.45%), Cyprus and Italy (+7.14%), and Hungary (+8.7%). Iceland, North Macedonia, and Serbia (−50%), Slovenia (−37.5%), Malta (−28.57%), Romania (−26.67%), and Bulgaria (−24.39%) registered the highest falls. In 2018 as opposed to 2016, North Macedonia (+200%), Turkey (+80%), Czechia (+72.22%), Belgium (+44.96%), and Hungary (+36%)

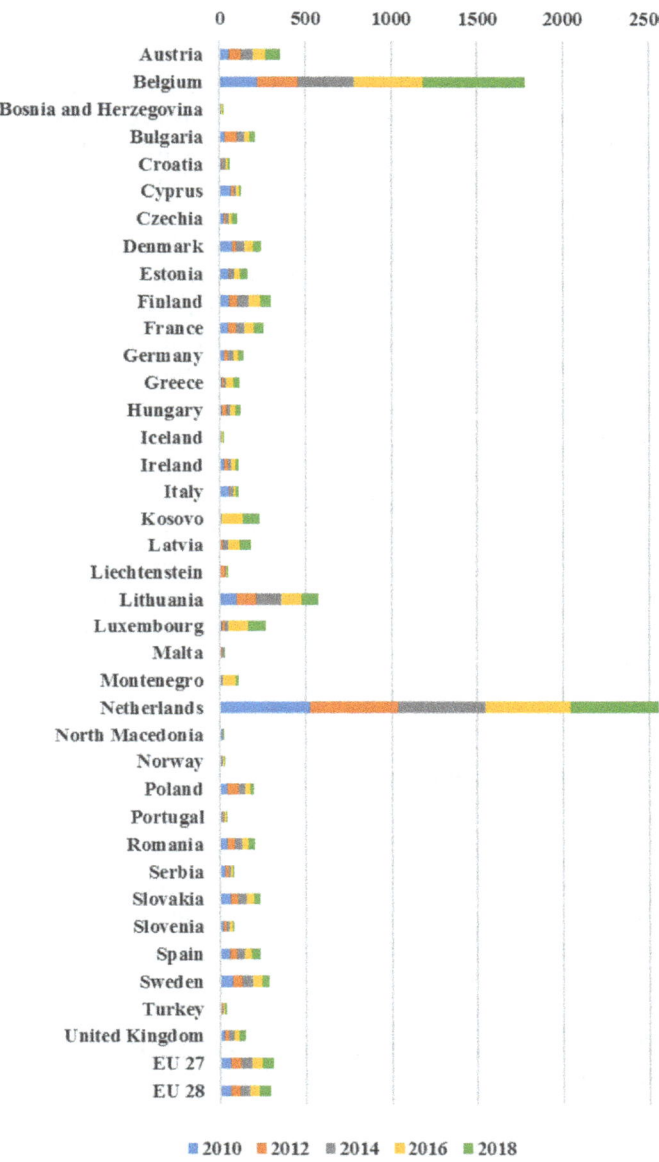

Fig. 6 Vegetal waste generated by economic activities in the European countries between 2010 and 2018 (kilograms per capita) (*Source* Author's development based on Eurostat [2021b])

recorded the highest increase, and Greece (+2.63%), the United Kingdom (+3.03%), France (+5.77%), Italy (+6.67%), and Bulgaria (+9.68%) registered the lowest expansions. At the opposite end, the following cases with the highest reductions can be emphasized: Bosnia and Herzegovina (−85.71%), Montenegro (−68.12%), Liechtenstein (−55.56%), Ireland (−40.74%), and Slovenia (−40%). Albania (2012/2010, 2014/2012, 2016/2014, and 2018/2016), Malta (2014/2012), Norway (2014/2012 and 2018/2016), and Finland (2018/2016) maintained a constant level of their waste.

The analysis emphasized the following: 5 (Austria, Belgium, Bulgaria, Kosovo, and Luxembourg) out of 38 countries registered both the highest level of waste and important boost as compared to previous year, 5 (Belgium, Bulgaria, Cyprus, Denmark, and Lithuania) out of 38 countries recorded both the highest level of waste and minor increase as compared to previous year, 4 (Bulgaria, Cyprus, Denmark, and Kosovo) out of 38 countries recorded both the highest level of waste and significant fall as against previous year, 8 (Croatia, Greece, Iceland, Kosovo, Latvia, Montenegro, North Macedonia, and Turkey) out of 38 countries recorded both the lowest level of waste and a significant increase versus previous year, 8 (Bosnia and Herzegovina, Iceland, Kosovo, Liechtenstein, Malta, Montenegro, North Macedonia, and Serbia) out of 38 countries recorded both the lowest level of waste and a major fall as opposed to previous year, and 3 (Bosnia and Herzegovina, Greece, and Malta) out of 38 countries recorded both the lowest level of waste and slight increase as opposed to previous year.

3.3 Analysis of Household and Similar Wastes

With respect to the household and similar wastes, Eurostat (2021b) reported data every two years between 2004 and 2018. To ensure data comparison with animal and mixed food waste and vegetal waste, the analysis will be conducted for the same period of time that was used for these two types of waste, i.e., 2010–2018.

3.3.1 Analysis of Household and Similar Wastes Generated by Households

In the case of household and similar wastes generated by households (which can be framed as food waste), in 2010, Portugal (470 kg per capita), Greece and Spain (413 kg per capita), Turkey (412 kg per capita), Italy (361 kg per capita), and Denmark (334 kg per capita) recorded the highest levels, and Cyprus (2 kg per capita), Estonia (122 kg per capita), Belgium (147 kg per capita), Finland (160 kg per capita), and Iceland (165 kg per capita) registered the lowest levels. Albania, Bosnia and Herzegovina, Croatia, Kosovo, Liechtenstein, Montenegro, and Serbia recorded null values or have not reported any data. In 2012, the highest levels were in the case of Portugal (413 kg per capita), Turkey (406 kg per capita), Spain (380 kg per capita), Cyprus (378 kg per capita), and Greece (377 kg per capita). By contrast, the countries with the lowest levels were Liechtenstein (84 kg per capita), Estonia (128 kg per capita), Belgium (132 kg per capita), Slovenia (174 kg per capita), and Finland (178 kg per capita). The following states recorded null values or have not reported any data: Albania, Bosnia and Herzegovina, Montenegro, North Macedonia, and Serbia (Fig. 7).

In 2014, Albania (425 kg per capita), Portugal (410 kg per capita), Turkey (394 kg per capita), Spain (365 kg per capita), and Greece (363 kg per capita) recorded the highest levels, and Slovenia (134 kg per capita), Finland (138 kg per capita), Estonia (153 kg per capita), Belgium (170 kg per capita), and Sweden (173 kg per capita) registered the lowest levels. There are only two states that recorded null values or have not reported any data, namely, Bosnia and Herzegovina and North Macedonia. In 2016, the top 5 countries with the highest levels included Portugal (427 kg per capita), Cyprus (404 kg per capita), Spain (388 kg per capita), Greece (381 kg per capita), and Turkey (334 kg per capita). Conversely, the states with the lowest levels were Slovenia (128 kg per capita), Estonia (147 kg per capita), Finland (162 kg per capita), Belgium (163 kg per capita), and Kosovo (170 kg per capita). Albania, Bosnia and Herzegovina, and North Macedonia recorded null values or have not reported any data.

In 2018, Portugal (435 kg per capita), Spain (391 kg per capita), Cyprus (384 kg per capita), Greece (381 kg per capita), and Montenegro (359 kg per capita) recorded the highest level, while Slovenia (129 kg per capita), Estonia (151 kg per capita), Belgium (158 kg per capita), Latvia (175 kg per capita), and Norway (174 kg per capita) registered the lowest levels. There are only two states that recorded null values or have not reported any data, i.e., Albania and North Macedonia.

Secondly, the evolution of the household and similar wastes generated by households showed that, in 2012 against 2010, Cyprus (+18,800%), Latvia (+60.96%), Norway (+16%), Finland (+11.25%), and Austria (+8.89%) recorded the highest rises, and Denmark (+0.6%), Turkey (+1%), Poland (+3.64%), Sweden (+4.78%), and Estonia (+4.92%) registered the lowest growths. The highest declines were in the case of North Macedonia (−100%), Slovenia (−29.55%), Bulgaria (−19.14%), Malta (−16.19%), and Romania (−15.45%). Albania, Bosnia and Herzegovina, Montenegro, and Serbia generated a similar level of waste. In 2014 as compared to 2012, there were only 8 states that recorded increases, Liechtenstein (+173.81%), Iceland (+98.03%), Belgium (+28.79%), Estonia (+19.53%), and Hungary (+18.18%) registered the highest increases and Austria (+0.51%), Slovakia (+1.18%), and Bulgaria (+4.58%) recorded the lowest boosts. The highest falls were for Latvia (−30.9%), Sweden (−28.22%), Slovenia (−22.99%), Finland (−22.47%), and Poland (−20.61%).

In 2016 versus 2014, the following cases with the highest increases can be underscored, namely Luxembourg (+25.89%), Montenegro (+25.61%), Cyprus (+25.08%), Finland (+17.39%), and Latvia (+16.35%), which come into contrast with the lowest increases in the United Kingdom (+0.44%), Slovakia (+0.78%), Romania (+1.14%), Serbia (+1.43%), and Sweden (+2.89%). At the opposite end, the highest drops were for Albania (−100%), Kosovo (−37.04%), Bulgaria (−17.15%), Turkey (−15.23%), and Lithuania (−9.38%). Bosnia and Herzegovina and North

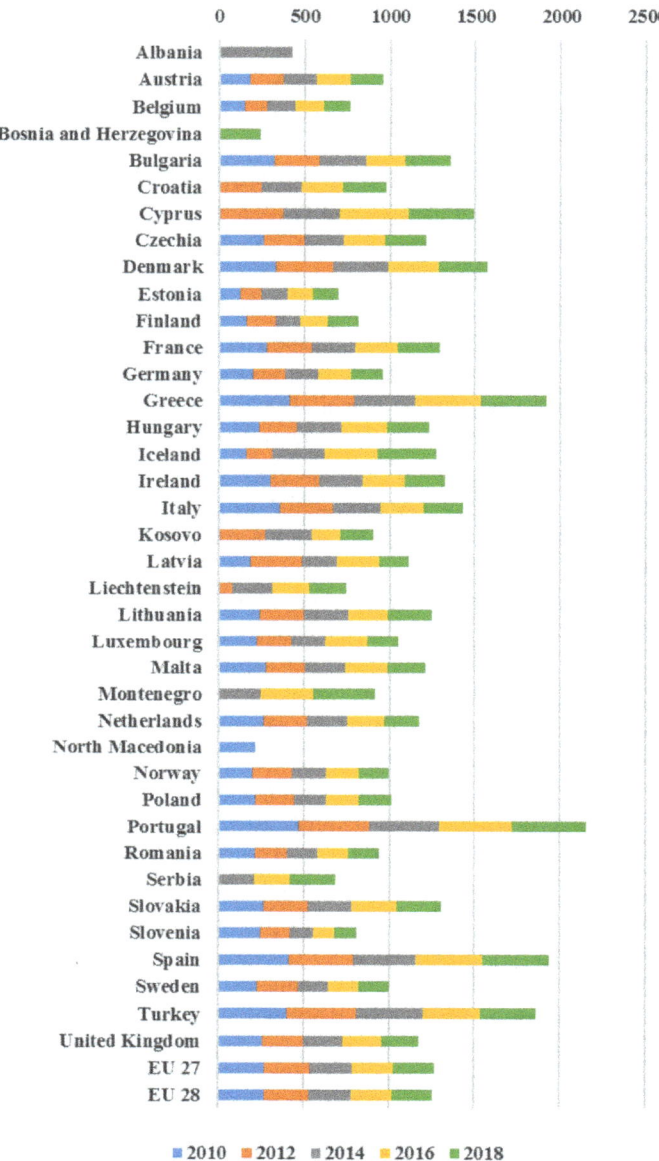

Fig. 7 Household and similar wastes generated by households in the European countries between 2010 and 2018 (kilograms per capita) (*Source* Author's development based on Eurostat [2021b])

Macedonia maintained a constant level of waste in 2016 and 2014 as compared to previous years. In 2018 in contrast to 2016, Serbia (+24.41%), Bulgaria (+18.06%), Montenegro (+16.18%), Finland (+11.73%), and Lithuania (+11.21%) recorded the highest booms, whereas Spain (+0.77%), Slovenia (+0.78%), Austria (+1.04%), Sweden (+1.69%), and Portugal (+1.87%) registered the lowest rises. The highest declines occurred in the case of Luxembourg (−29.03%), Latvia (−27.69%), Norway (−9.84%), Italy (−9.09%), and Malta (−8.57%). Albania and North Macedonia provided a similar level of waste.

The analysis underlined that 2 (Cyprus and Montenegro) out of 38 countries recorded both the highest level of waste and major boost as compared to previous year, 6 (Austria, Denmark, France, Germany, Luxembourg, and the United Kingdom) out of 38 countries recorded both the highest level of waste and minor increase as compared to previous year, 3 (Albania, Italy, and Turkey) out of 38 countries recorded both the highest level of waste and decline against previous year, 8 (Belgium, Cyprus, Estonia, Finland, Iceland, Latvia, Liechtenstein, and Norway) out of 38 countries recorded both the lowest level of waste and a significant increase versus previous year, 3 (Estonia, Slovenia, and Sweden) out of 38 countries recorded both the lowest level of waste and minor increase as opposed to previous year, and 6 (Finland, Kosovo, Latvia, Norway, Slovenia, and Sweden) out of 38 countries recorded both the lowest level of waste and decline as compared to the previous year.

3.3.2 Analysis of Household and Similar Wastes Generated by Economic Activities

As regards the household and similar wastes generated by economic activities (which can be enclosed as food loss), in 2010, the higher levels were for Ireland (413 kg per capita), Croatia (353 kg per capita), Austria (258 kg per capita), Malta (248 kg per capita), and the United Kingdom (235 kg per capita), while the lowest levels were for North Macedonia, Norway, and Turkey (2 kg per capita), Serbia (3 kg per capita), Poland (7 kg per capita), Portugal (10 kg per capita), and Greece (16 kg per capita). Albania, Bosnia and Herzegovina, Italy, Kosovo, Montenegro, and Slovakia recorded null values or have not reported any data (Fig. 8). In 2012, Malta (257 kg per capita), Belgium (212 kg per capita), the United Kingdom (204 kg per capita), Luxembourg (184 kg per capita), and the Netherlands (176 kg per capita) recorded the highest levels, and states such as North Macedonia (1 kg per capita), Montenegro (2 kg per capita), Poland (3 kg per capita), Bosnia and Herzegovina (4 kg per capita), and Turkey (5 kg per capita) recorded the lowest levels. The countries that recorded null values or have not reported any data were Albania, Italy, Norway, and Slovakia.

In 2014, the following cases with the highest levels can be highlighted: Bosnia and Herzegovina (315 kg per capita), Malta (244 kg per capita), the United Kingdom (203 kg per capita), Belgium (197 kg per capita), and Luxembourg (185 kg per capita). By contrast, Italy (1 kg per capita), North Macedonia (2 kg per capita), Turkey (6 kg per capita), Greece (12 kg per capita), and Serbia (15 kg per capita) registered the lowest levels. In 2016, Bosnia and Herzegovina (328 kg per capita),

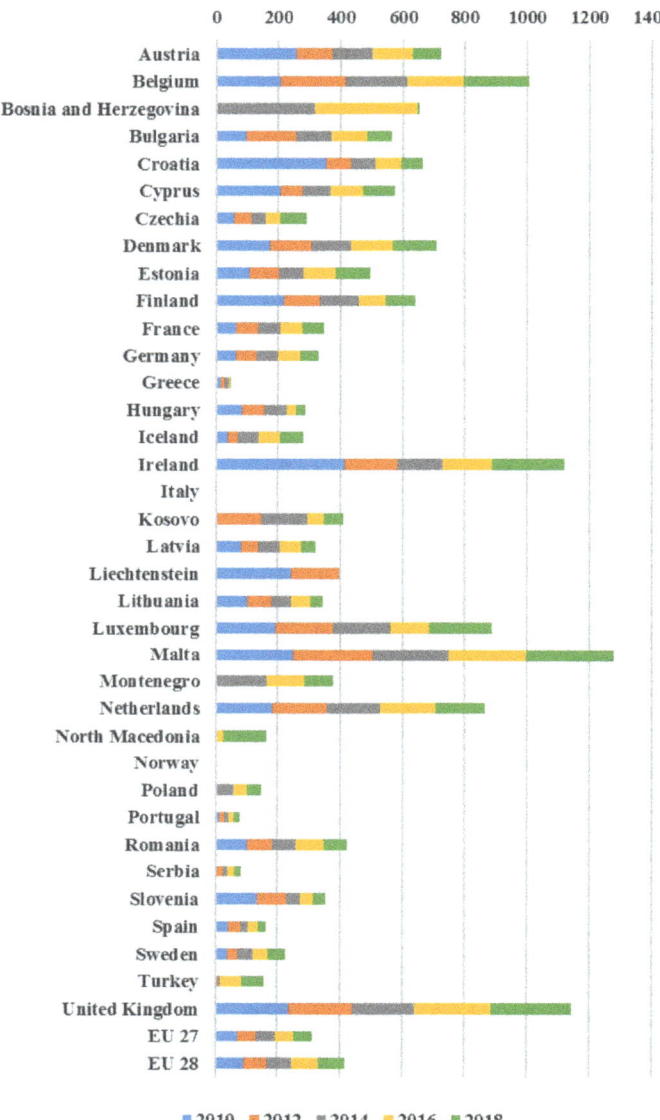

Fig. 8 Household and similar wastes generated by economic activities in the European countries between 2010 and 2018 (kilograms per capita) (*Source* Author's development based on Eurostat [2021b])

Malta (246 kg per capita), the United Kingdom (243 kg per capita), Belgium (180 kg per capita), and the Netherlands (178 kg per capita) recorded the highest levels. The lowest levels were in the case of Italy (1 kg per capita), Greece (2 kg per capita), Portugal (15 kg per capita), North Macedonia (20 kg per capita), and Serbia (22 kg per capita).

In 2018, the highest levels were for Malta (283 kg per capita), the United Kingdom (261 kg per capita), Ireland (235 kg per capita), Belgium (211 kg per capita), and Luxembourg (198 kg per capita), and the lowest levels were for Italy (2 kg per capita), Greece (4 kg per capita), Bosnia and Herzegovina (8 kg per capita), Portugal and Serbia (21 kg per capita), and Spain (26 kg per capita). Nonetheless, Albania, Liechtenstein, Norway, Slovakia recorded null values or have not reported any data in 2014, 2016, and 2018.

Secondly, the evolution of the household and similar wastes generated by economic activities underlined that, in 2012 as compared to 2010, there were only 8 states that recorded increases, the highest occurred in the case of Serbia (+600%), Turkey (+150%), Bulgaria (+70.83%), Portugal (+70%), and France (+12.5%), whereas the lowest were registered by Belgium (+3.41%), Malta (+3.63%), and Germany (+4.69%). The highest declines were reported by Norway (−100%), Croatia (−77.05%), Cyprus (−64.56%), Ireland (−58.11%), and Poland (−57.14%). In 2014 against 2012, Montenegro (+8,000%), Bosnia and Herzegovina (+7,775%), Poland (+1,466.67%), North Macedonia (+100%), and Iceland (+94.12%) recorded the highest rises, while Luxemburg (+0.54%), Kosovo (+0.68%), Hungary (+2.78%), Germany (+2.99%), and Finland (+6.9%) registered the lowest boosts. The highest falls were for Liechtenstein (−100%), Slovenia (−59.18%), Spain (−33.33%), Bulgaria (−32.93%), and Lithuania (−21.52%).

In 2016 in contrast to 2014, the highest growths were in the case of Turkey (+1,100%), North Macedonia (+900%), Serbia (+46.67%), Estonia (+30%), and Cyprus (+20.45%), and the lowest increases were for Malta (+0.82%), Germany (+1.45%), Austria (+2.34%), Croatia (+2.53%), and Lithuania (+3.23%). Greece (−83.33%), Kosovo (−63.27%), Hungary (−59.46%), Luxembourg (−31.35%), and Finland (−30.65%) recorded the highest drops. Finally, in 2018 as opposed to 2016, the top 5 countries with the highest increases included: North Macedonia and Turkey (+585%), Greece and Italy (+100%), Czechia (+93.33%), Luxembourg (+55.91%), and Portugal (+40%). Countries such as Sweden (+1.92%), Demark (+5.97%), Estonia (+6.73%), the United Kingdom (+.41%), and Iceland (+8.7%) registered the lowest booms. The highest decreases were in the case of Bosnia and Herzegovina (−97.56%), Lithuania (−40.63%), Austria and Bulgaria (−31.3%), Latvia (−29.41%), and Montenegro (−23.14%).

Albania and Slovakia (2012/2010, 2014/2012, 2016/2014, and 2018/2016), Italy (2012/2010 and 2016/2014), Liechtenstein (2016/2014 and 2018/2016), Norway (2014/2012, 2016/2014, and 2018/2016), and Portugal (2016/2014) generated similar levels of waste.

The analysis stressed that 2 (Bosnia and Herzegovina and Luxembourg) out of 38 countries recorded both the highest level of waste and important increase as compared to previous year, 6 (Austria, Belgium, Croatia, Luxembourg, Malta, and the United Kingdom) out of 38 countries recorded both the highest level of waste and minor rise as compared to previous year, 5 (Austria, Bosnia and Herzegovina, Croatia, Ireland, and Luxembourg) out of 38 countries recorded both the highest level of waste and drop against previous year, 5 (Bosnia and Herzegovina, Greece, Italy, Montenegro, and Turkey) out of 38 countries recorded both the lowest level of waste and a major increase versus previous year, no countries recorded both the lowest level of waste and minor boost as opposed to previous year, and 6 (Bosnia and Herzegovina, Greece, Montenegro, Norway, Poland, and Spain) out of 38 countries recorded both the lowest level of waste and major decrease as compared to previous year.

4   Conclusions

Currently, the Eurostat database is the only framework that contains food waste data which can be analyzed. But, in the following years, as FAO will collect more data and will establish a time series of FLI and FWI, the FLW Database will become more useful for researchers.

By analyzing the animal and mixed food waste, vegetal waste, and household and similar wastes generated by households, the following cluster of countries with similar behavior can be stressed (disregarding the countries that reported null values or have not reported any data). The ranking on "high" and "low" was made based on the position of each country's food waste level against the average value of all countries' level for each year. Thus, Italy registered high levels of all three waste categories; Bosnia and Herzegovina, Malta, and Serbia registered low levels of all three waste categories; Finland and the Netherlands recorded high levels of animal and mixed food waste and low levels of vegetal waste and household and similar wastes; Austria, Luxemburg, Norway, and Sweden recorded high levels of animal and mixed food waste and vegetal waste and low levels of household and similar wastes; Belgium and Slovenia recorded high levels of vegetal waste and low levels of animal and mixed food waste and household and similar wastes; Denmark, Iceland, and the United Kingdom recorded high levels of vegetal waste and household and similar wastes and low levels of animal and mixed food waste; Bulgaria, Croatia, Czechia, Hungary, Ireland, Montenegro, Slovakia, and Spain recorded high levels of household and similar wastes and low levels of animal and mixed food waste and vegetal waste.

Most of the countries with the highest level of the three waste categories generated by households recorded the lowest change (increase or decline) between 2010 and 2018. This is the case of Austria, Denmark, Germany, Greece, Italy, Liechtenstein, the Netherlands, Norway, Portugal, Spain, etc. with some exceptions in specific years, such as Belgium, Cyprus, Finland, and Luxembourg. The vice versa is applicable too, for instance, Bulgaria, Estonia, Latvia, Iceland, Ireland, Poland, Romania, Serbia, Slovakia, Slovenia, apart from Sweden, the United Kingdom, etc.

In the case of animal and mixed food waste, vegetal waste, and household and similar wastes generated by economic activities, the following cluster of states with similar behavior can be underscored (the states that reported null values or have not reported any data are not included): Austria, Belgium, Finland, and the Netherlands registered high levels of all three waste categories; Bosnia and Herzegovina, Croatia, Czechia, Germany, Greece, Hungary, Italy, Kosovo, Latvia, North Macedonia, Poland, Portugal, Romania, Serbia, Slovenia, Spain, and Turkey registered low levels of all three waste categories; France, Iceland, Liechtenstein, and Norway recorded high levels of animal and mixed food waste and low levels of vegetal waste and household and similar wastes; Cyprus, Ireland, and the United Kingdom recorded high levels of animal and mixed food waste and household and similar wastes and low levels of vegetal waste; Sweden recorded high levels of animal and mixed food waste and vegetal waste and low levels of household and similar wastes; Lithuania recorded high levels of vegetal waste and low levels of animal and mixed food waste and household and similar wastes; Bulgaria, Denmark, Estonia, Luxembourg, Malta, and Montenegro recorded high levels of household and similar wastes and low levels of animal and mixed food waste and vegetal waste.

The same pattern of the negative relationship between the country's waste level and the weight of the changes in the country's waste level was applicable for the animal and mixed food waste, vegetal waste, and household and similar wastes generated by economic activities for 2010–2018 period. Hence, Belgium, Norway, Lithuania, Luxembourg, Malta, the Netherlands, the United Kingdom, etc. recorded high level of waste and low changes in their level. Finland, France, Iceland, and Ireland were special cases that have disobeyed for certain years the above-mentioned pattern.

These findings substantiate a positive correlation between the country's economic development and the quantity of food waste generated by households and economic activities.

As regards the limitations of this research, one can notice from the outset that the data of the animal and mixed food waste, the vegetable wastes and the household and similar wastes contain data of other type of wastes in addition to food waste and thus, a clear distinction between them cannot be made. Secondly, the analysis was constrained by the fact that, between 2004 and 2018, particular countries have not reported any data to Eurostat for important food waste category (for instance, Albania, Cyprus, France, Germany, Greece, Kosovo, Latvia, Liechtenstein, Lithuania, North Macedonia, Poland, Portugal, Romania, and Turkey in the case of the animal and mixed food waste generated by households), individual states reported erroneous data (such as Montenegro in 2012 for household and similar waste generated by economic activities, Kosovo in 2012 and 2014 for vegetal wastes generated by economic activities, etc.), whereas certain countries stopped reporting data (for example Norway between 2012 and 2018 in the case of household and similar waste generated by economic activities).

References

Alabrese, M., Brunori, M., Rolandi, S., & Saba, A. (2015). Agri-food industries and the challenge of reducing food wastage: An analysis of legal opportunities. In L. Escajedo San-Epifanio & M. de Renobales Scheifler (Eds.), *Envisioning a future without food waste and food poverty: Societal challenges*. Wageningen Academic Publishers, Wageningen. https://doi.org/10.3920/978-90-8686-820-9

Blakeney, M. (2019). *Food loss and food waste: Causes and solutions*. Edward Elgar Publishing.

Corvellec, H. (2013). Recycling food waste into biogas, or how management transforms overflows into flows. In B. Czarniawska & O. Löfgren (Eds.), *Coping with excess: How organizations, communities and individuals manage overflows*. Edward Elgar Publishing.

Ene, C. (2013). Post-consumer waste—Challenges, trends, solutions. *International Journal of Sustainable Economies Management, 2*(3), 19–31. https://www.igi-global.com/article/post-consumer-waste/94586

Ene, C., Voica, M. C., & Panait, M. (2017). Green investments and food security: Opportunities and future directions in the context of sustainable development. In M. Mieila (Ed.), *Measuring sustainable development and green investments in contemporary economies*. IGI Global, Hershey. https://doi.org/10.4018/978-1-5225-2081-8

European Commission. (2002). *Regulation No 2150 of the European Parliament and of the Council on waste statistics*. https://eur-lex.europa.eu/legal-content/en/ALL/?uri=CELEX%3A32002R2150. Accessed 20 July 2021.

European Commission. (2008). *Directive 98/EC of the European Parliament and of the Council of 19 November 2008 on waste and repealing certain directives*. https://eur-lex.europa.eu/legal-content/EN/TXT/?uri=celex%3A32008L0098. Accessed 20 July 2021.

European Commission. (2011). *Decision 753/EU: Commission decision of 18 November 2011 establishing rules and calculation methods for verifying compliance with the targets set in Article 11(2) of Directive 2008/98/EC of the European Parliament and of the Council*. https://eur-lex.europa.eu/legal-content/EN/TXT/?uri=CELEX%3A02008L0098-20180705. Accessed 20 July 2021.

European Commission. (2019a). *Commission delegated decision (EU) 1597 of 3 May 2019 supplementing Directive 2008/98/EC of the European Parliament and of the Council as regards a common methodology and minimum quality requirements for the uniform measurement of levels of food waste*. https://eur-lex.europa.eu/legal-content/EN/TXT/?uri=uriserv%3AOJ.L_.2019.248.01.0077.01.ENG. Accessed 21 July 2021.

European Commission. (2019b). *Commission implementing decision (EU) 2000 of 28 November 2019 laying down a format for reporting of data on food waste and for submission of the quality check report in accordance with Directive 2008/98/EC of the European Parliament and of the Council*. https://eur-lex.europa.eu/legal-content/EN/TXT/?uri=uriserv%3AOJ.L_.2019.310.01.0039.01.ENG&toc=OJ%3AL%3A2019%3A310%3ATOC. Accessed 21 July 2021.

Eurostat. (2013). *Manual on waste statistics—A handbook for data collection on waste generation and treatment*. https://ec.europa.eu/eurostat/documents/3859598/5926045/KS-RA-13-015-EN.PDF.pdf/055ad62c-347b-4315-9faa-0a1ebcb1313e?t=1414782620000. Accessed 21 July 2021.

Eurostat. (2021a). *Guidance on reporting of data on food waste and food waste prevention according to Commission implementing decision (EU) 2019/2000*. https://ec.europa.eu/eurostat/documents/342366/351811/Guidance+on+food+waste+reporting/5581b0a2-b09e-adc0-4e0a-b20062dfe564. Accessed 21 July 2021.

Eurostat. (2021b). *Generation of waste by waste category, hazardousness and NACE Rev. 2 activity [env_wasgen]*. https://ec.europa.eu/eurostat/web/waste/data/database. Accessed 21 July 2021.

Eurostat. (2021c). *Information. Generation of waste by waste category, hazardousness and NACE Rev. 2 activity [env_wasgen]*. https://ec.europa.eu/eurostat/estat-navtree-portlet-prod/NodeInfoServices?lang=en&code=env_wasgen. Accessed 27 July 2021.

Fabi, C. (2020). *SDG 12.3.1.a. Food loss index*. http://www.fao.org/3/cb2949en/cb2949en.pdf. Accessed 23 July 2021.

FAO, IFAD, UNICEF, WFP, WHO. (2021). The state of food security and nutrition in the world 2021. *Transforming food systems for food security, improved nutrition and affordable healthy diets for all.* https://doi.org/10.4060/cb4474en

Food and Agriculture Organization of the United Nations. (2011). *Global food losses and food waste: Extent, causes and prevention.* http://www.fao.org/3/i2697e/i2697e.pdf. Accessed 16 July 2021.

Food and Agriculture Organization of the United Nations. (2015). *Food loss and waste facts.* http://www.fao.org/resources/infographics/infographics-details/en/c/317265/. Accessed 14 July 2021.

Food and Agriculture Organization of the United Nations. (2017). *The future of food and agriculture: Trends and challenges.* Food and Agriculture Organization of the United Nations.

Food and Agriculture Organization of the United Nations. (2019a). *The State of Food and Agriculture 2019: Moving forward on food loss and waste reduction.* http://www.fao.org/3/ca6030en/ca6 030en.pdf. Accessed 16 July 2021.

Food and Agriculture Organization of the United Nations. (2019b). *Metadata of indicator 12.3.1.* Global Food Loss Index. http://www.fao.org/3/CA2593EN/ca2593en.pdf. Accessed 16 July 2021.

Food and Agriculture Organization of the United Nations. (2019c). *Methodological proposal for monitoring SDG target 12.3.1 sub-indicator 12.3.1.a.* The Food loss index design, data collection methods and challenges. http://www.fao.org/3/ca4012en/ca4012en.pdf. Accessed 23 July 2021.

Food and Agriculture Organization of the United Nations. (2021a). *Food loss and waste database.* http://www.fao.org/platform-food-loss-waste/flw-data/en/. Accessed 23 July 2021.

Food and Agriculture Organization of the United Nations. (2021b). *How to use the FLW database.* http://www.fao.org/platform-food-loss-waste/flw-data/user-guide/en/. Accessed 23 July 2021.

Food and Agriculture Organization of the United Nations. (2021c). *Sustainable development goals: Indicator 12.3.1—global food loss and waste.* http://www.fao.org/sustainable-develo pment-goals/indicators/12.3.1/en/. Accessed 23 July 2021.

Food and Agriculture Organization of the United Nations. (2021d). *SDG 12.3.1: Global food loss index.* http://www.fao.org/3/CA2640EN/ca2640en.pdf. Accessed 23 July 2021.

Food and Agriculture Organization of the United Nations. (2021e). *Technical platform on the measurement and reduction of food loss and waste.* http://www.fao.org/platform-food-loss-waste/ food-loss/food-loss-measurement/en/. Accessed 23 July 2021.

Food and Agriculture Organization of the United Nations. (2021f). *Food and agriculture statistics.* http://www.fao.org/food-agriculture-statistics/capacity-development/food-loss-and-waste/en/. Accessed 23 July 2021.

Garnsey, P. (2004). *Cities, peasants and food in classical antiquity: Essays in social and economic history.* Cambridge University Press.

Gennari, P. (2015). *SDG target 12.3. indicator 12.3.1—Global food loss index Food and Agriculture Organization of the United Nations.* https://unstats.un.org/sdgs/files/meetings/iaeg-sdgs-meeting-03/3rd-IAEG-SDGs-presentation-FAO--12.3.1.pdf. Accessed 23 July 2021.

Hartikainen, H., Svanes, E., Franke, U., Mogensen, L., Andersen, S., Bond, R., Burman, C., Einarsson, E., Joensuu, P. E. K., Olsson, M. E., Räikkönen, R., Sinkko, T., Stubhaug, E., Rosell, A., & Sundin, S. (2017). *Food losses and waste in primary production: Case studies on carrots, onions, peas, cereals and farmed fish.* Nordic Council of Ministers.

High Level Panel of Experts on Food Security and Nutrition. (2014). *Food losses and waste in the context of sustainable food systems.* http://www.fao.org/3/i3901e/i3901e.pdf. Accessed 17 July 2021

Horton, P., Bruce, R., Reynolds, C., & Milligan, G. (2019). Food Chain Inefficiency (FCI): Accounting conversion efficiencies across entire food supply chains to re-define food loss and waste. *Frontiers in Sustainable Food Systems, 3,* 31–41. https://doi.org/10.3389/fsufs.2019.00079

Kummu, M., de Moel, H., Porkka, M., Siebert, S., Varis, O., & Ward, P. J. (2012). Lost food, wasted resources: Global food supply chain losses and their impacts on freshwater, cropland and fertilizer use. *Science of Total Environment, 438,* 477–489. https://doi.org/10.1016/j.scitotenv.2012.08.092

Lagioia, G., Amicarelli, V., Gallucci, T., & Bux, C. (2021). Food waste reduction towards food sector sustainability. In: Information Resources Management Association (Ed.), *Research anthology on food waste reduction and alternative diets for food and nutrition security*. IGI Global, Hershey. doi:https://doi.org/10.4018/978-1-7998-5354-1.ch027

McDermott, B. (2001). *Decoding Egyptian hieroglyphs: How to read the secret language of the Pharaohs*. Chronicle Books.

Parfitt, J., Barthel, M., & Macnaughton, S. (2010). Food waste within food supply chains: Quantification and potential for change to 2050. *Philosophical Transactions of the Royal Society B, 365*, 3065–3081. https://doi.org/10.1098/rstb.2010.0126

Pinstrup-Andersen, P., Gitz, V., & Meybeck, A. (2016). Food losses and waste and the debate on food and nutrition security. In B. Pritchard, R. Ortiz, & M. Shekar (Eds.), *Routledge handbook of food and nutrition security*. Routledge.

Schneider, F. (2013). Review of food waste prevention on an international level. *Proceedings of the Institution of Civil Engineers, Waste and Resource Management, 166*, 187–203. https://doi.org/10.1680/warm.13.00016

Smil, V. (2004). Improving efficiency and reducing waste in our food system. *Environment Science, 1*, 17–26. https://doi.org/10.1076/evms.1.1.17.23766

The World Bank. (2020). *Addressing food loss and waste: A global problem with local solutions*. https://openknowledge.worldbank.org/bitstream/handle/10986/34521/Addressing-Food-Loss-and-Waste-A-Global-Problem-with-Local-Solutions.pdf?sequence=1&isAllowed=y. Accessed 16 July 2021.

United Nations. (2015). Resolution adopted by the General Assembly on 25 September 2015. *Transforming our world: The 2030 agenda for sustainable development*. https://www.un.org/ga/search/view_doc.asp?symbol=A/RES/70/1&Lang=E. Accessed 18 July 2021.

United Nations. (2021). *Sustainable development goals*. http://www.fao.org/sustainable-development-goals/overview/en/. Accessed 14 July 2021.

United Nations Environment Programme. (2021). *UNEP food waste index report*. https://www.unep.org/resources/report/unep-food-waste-index-report-2021. Accessed 23 July 2021.

United States Environmental Protection Agency. (2014). *Municipal solid waste generation, recycling, and disposal in the United States tables and figures for 2012*. https://www.epa.gov/sites/default/files/2015-09/documents/2012_msw_dat_tbls.pdf. Accessed 22 July 2021.

United States Environmental Protection Agency. (2020a). *Wasted food measurement methodology scoping memo*. https://www.epa.gov/sites/default/files/2020-06/documents/food_measurement_methodology_scoping_memo-6-18-20.pdf. Accessed 22 July 2021.

United States Environmental Protection Agency. (2020b). *2018 wasted food report*. https://www.epa.gov/sites/default/files/2020-11/documents/2018_wasted_food_report.pdf. Accessed 22 July 2021.

von Grebmer, K., Bernstein, J., Alders, R., Dar, O., Kock, R., Rampa, F., Wiemers, M., Acheampong, K., Hanano, A., Higgins, B., Ní Chéilleachair, R., Foley, C., Gitter, S., Ekstrom, K., & Fritschel, H. (2020). *2020 Global hunger index: One decade to zero hunger: Linking health and sustainable food systems*. https://www.globalhungerindex.org/pdf/en/2020.pdf. Accessed 14 July 2021.

World Economic Forum. (2018). *System initiative on shaping the future of food security and agriculture innovation with a purpose: The role of technology innovation in accelerating food systems transformation*. http://www3.weforum.org/docs/WEF_Innovation_with_a_Purpose_VF-reduced.pdf. Accessed 16 July 2021.

Sustainability, Innovation and Diversification in the Spanish Frozen Food Industry: A Financial Analysis

Félix Puime-Guillén, Raisa Pérez-Vas, and Raquel Fernández-González

Abstract Spain is the leading European country in the production and consumption of frozen products. More specifically, in the region of Galicia (northwest of the Iberian Peninsula) the bulk of the frozen fish and seafood industry is located. The production activity carried out in Spain, as a member of the European Union, must be governed by European sustainability guidelines. This is what one of the largest business groups in the frozen sector in Galicia has done in its new business project, a plant producing high-quality frozen products with a significant added value. The objective of this study is to analyze the viability of this new company, over a five-year financial projection. In addition, a price sensitivity analysis will also be carried out in order to determine the possible scenarios that the company would face. The results show that the project is efficient and covers the investments made, in the medium term, despite the losses suffered in the short term.

Keywords Sustainability · Innovation · Diversification · Spain · Frozen food industry · Financial analysis

1 Introduction

In 2019, the frozen food market reached €285 million, and this figure is expected to be even higher in 2020 (Grand View research, 2021). The growing trend in this sector is the result of increasing consumer confidence in the product but, especially, due to new occidental eating habits, where celerity and availability become essential (Carlson &

F. Puime-Guillén (✉)
Department of Business, University of A Coruña, A Coruña, Spain
e-mail: felix.puime@udc.es

R. Pérez-Vas
Department of Financial Economics and Accounting, IC2-ECOBAS, University of Vigo, Vigo, Spain
e-mail: raiperez@uvigo.es

R. Fernández-González
Department of Applied Economics, ERENEA-ECOBAS, University of Vigo, Vigo, Spain
e-mail: raquelf@uvigo.es

© The Author(s), under exclusive license to Springer Nature Switzerland AG 2022   149
C. Machado and J. P. Davim (eds.), *Sustainability and Intelligent Management*,
Management and Industrial Engineering, https://doi.org/10.1007/978-3-030-98036-8_7

Frazão, 2014; Nussinovitch & Peleg, 2000; Sen et al., 2021). Accordingly, global estimated turnover for the frozen food industry is expected to reach €345 million in 2017 (Business Wire, 2021). Therefore, the frozen food sector is an industry oriented toward the future.

However, it is imperative that the growth process of the sector be carried out in a sustainable manner. One of the Sustainable Development Goals (SDGs), adopted by the United Nations in 2015, is the 12th goal: "Ensure sustainable consumption and production patterns". It advocates increasing efficiency in both food production and consumption, while reducing environmental degradation and ensuring a more equitable distribution of food resources (United Nations, 2021). For this purpose, it is necessary to focus on innovation, both in the design of new products and in the different links that compose the production chain, such as distribution, storage and packaging (León-Bravo et al., 2019; Yakovleva & Flynn, 2004).

In the fish and seafood processing industry, innovation in the freezing process is of great importance. Fish is a food source containing essential nutrients, which contributes beneficially to human health (Chen et al., 2022; Li et al., 2020). However, fish and seafood have a fast decomposition process, which deteriorates their organoleptic properties and makes them unviable for consumption (Semeano et al., 2018). Therefore, freezing is an indispensable process in the fish and shellfish industry, as it prolongs the product storage life (Fuentes et al., 2013).

The European Union is one of the leading regions in frozen seafood consumption, but it is also one of the regions with the strongest food safety policies and a very strong commitment to sustainability in business (D.'Amico et al., 2018). Spain, a member country of the European Union, is the largest producer and consumer of frozen fish and seafood products in Europe. Precisely, the region of Galicia (located in the northwest of Spain) is where the largest number of companies, and the most important business groups, in the sector are located (FAO, 2021).

The objective of this study is to analyze, from an economic and financial perspective, the evolution of a company dependent on one of the large business groups in the frozen food sector in Spain. This new company is focused on the innovation of frozen products, offering high value-added products, through a sustainable process. To this end, a five-year financial valuation of the company will be carried out and, subsequently, a price sensitivity analysis will be performed to consider the evolution of the company under different commercial scenarios.

2   Case Study

The case study of this analysis is located in Galicia (Fig. 1), a Spanish region with a strong fishing and seafood tradition (Garza-Gil et al., 2009). The biophysical characteristics (1,200 km of coastline) together with the high quality of the waters (upwelling) of this region have encouraged the consolidation of a prominent maritime-fishing industry that has evolved toward activities directly or indirectly

Fig. 1 Location of the Spanish region of Galicia

related to the fishing sector (Bode et al., 2009; Fernández-González et al., 2020, 2021).

One of these Galician industries linked to the fishing sector is the frozen sea products sector. Initially, most of the raw material used in this sector was local product, but in the last decades there has been a prevalence of product from distant fishing fleets and high seas fishing (Garza-Gil et al., 2017). As a consequence, one of the ports of Galicia, the one located in the city of Vigo, is the main port in frozen sea products trade in Spain (Autoridad portuaria de Vigo, 2021). Overall, in 2020, 569,919 tons of frozen seafood products were loaded and unloaded at the Port of Vigo, a year-on-year growth of 1.16% compared to 2019 (563,375 tons) (Industrias Pesqueras, 2021).

As shown in Table 1, Galicia is the leading Spanish region in the frozen sea products sector, a position it has held since the consolidation of the sector at the national level (Bello Bugallo et al., 2013). Catalonia, the region in second position, does not represent a threat to the leadership of Galicia (18.09% of the national turnover). Of the top 15 companies with the largest market share, 8 are based in Galicia. One of these large Galician companies is the focus of the economic-financial analysis carried out in this study.

The company under analysis is part of a holding linked to the fishing sector. The firm was founded in 2018 and its main activity is the production of high value-added precooked food products. The objectives that have led to the creation of this dependent company are several. Firstly, the parent company seeks to strengthen its leadership in the sector and, secondly, it also aims to diversify the range and quality of the products it markets in order to increase the profit margin. This diversification strategy has received strong institutional backing from the regional government of

Table 1 Overview of the frozen sector in Galicia and Spain in 2020

			Galicia	Spain
Companies	Total	n°	102	240
		%	42.50%	
	Industrial companies	n°	42	109
		%	38.53%	
	Marketing companies	n°	60	131
		%	45.80%	
Turnover	Total	mill. €	4,256.9	8,656.5
		%	49.18%	
	Industrial companies	mill. €	2,710.9	5,448.6
		%	49.75%	
	Marketing companies	mill. €	1,546.0	3,207.9
		%	48.19%	
Employment	Total	n°	5,395	14,255
		%	37.85%	
	Industrial companies	n°	4,290	10,732
		%	39.97%	
	Marketing companies	n°	1,105	3,523
		%	31.37%	

Source Own elaboration based on Ardan (2020)

Galicia, which has subsidized this project with a grant of €64,000. The justification presented by the Galician regional government for the provision of public resources to this private company is the high level of R&D&I required for the manufacture of this type of product and the commitment to a more sustainable production process based on the philosophy of the circular economy (La Voz de Galicia, 2021).

3   Methodology

The present study involves an analysis of the viability of a company operating in the frozen food sector. Innovation and new technologies lead to the production of more elaborated and high-quality products originating from the sea. Therefore, this company has decided to implement a plant to produce gourmet products.

The economic and financial data of the company were obtained from the Orbis Europe database belonging to the Bureau Van Dijk Electronic Publishing group (Bureau Van Dijk Electronic Publishing group, 2021). The search for the information on the company was carried out by means of the NIF code, which unequivocally identifies the firm and, once located, the variables necessary to carry out the analysis were selected.

With the purpose of analyzing the profitability of this investment, two valuation models based on discounted cash flows, the Net Present Value (NPV) and the Internal Rate of Return (IRR), will be applied. For this purpose, an estimate is made of both the free cash flows, which represent the cash flow generated by the company without taking into account its indebtedness, and the free cash flow for the shareholder, which values the cash flows generated by the company considering the leverage of the firm (Trippi, 1989). The free cash flows for the shareholder are the flows obtained by the company after assuming its debt commitments.

In addition, a sensitivity analysis will be performed to determine the effect of sales growth on the valuation of the business. This evaluation is a way of incorporating risk into project analysis (Borgonovo et al., 2010; Chakrabarti & Zeaiter, 2014).

3.1 Net Present Value

The NPV is a model for investment valuation and financial analysis of companies based on discounted cash flows. This model allows to know what the profit or loss of the company will be, based on the present value of free cash flows (Žižlavský, 2014). In order to know the present value of the cash flows, it is necessary to discount them at a discount rate, using the Weighted Average Cost of Capital (WACC) for the free cash flow and the cost of equity for the cash flow of the shareholders (Block, 2011; Vartiainen et al., 2020). Through this model the flows of receipts and payments are discounted at a discount rate in order to obtain the profitability of the investment (Dusseault & Pasquier, 2021; Peymankar et al., 2021). Mathematically:

$$NPV = -I + \sum_{t=1}^{T} \frac{CF_t}{(1+i)^t} \qquad (1)$$

It is necessary for the NPV of a project to be greater than 0 in order for it to be carried out, so that it will provide future benefits. If the NPV is equal to 0, the project will generate neither profits nor losses, and if it is less than 0, the project will generate losses.

3.2 Interest Rate of Return

The IRR is a model widely used in business analysis (Arrow & Kruz, 2013). This model measures the interest or profitability provided by an investment. This model is closely linked to NPV, as it provides the discount rate that makes NPV equal to zero. Mathematically:

$$0 = -I + \sum_{t=1}^{T} \frac{CF_t}{(1+r)^t} \tag{2}$$

4   Results

The first step in calculating the viability of the company using the two models described above is to make a projection of the Profit and Loss Statement (Table 2). Through this projection the cash flows have been calculated. The discount rate used is 1.5%. In addition, a continuation cash flow of the company has been calculated for the last year under study.

As shown in Table 3, the cash flows have a negative balance for the first half of the period studied, while the values become positive in the last years analyzed. However, it should be noted that this positive data can be misleading, since positive balances can be obtained if the company gets increasingly indebted and covers the interest on the debt with this indebtedness. This is not the case in this study, in period 4, after

Table 2 Cash-flows projection

	2019	2020	2021	2022	2023
Revenue	**239,181.00**	**478,362.00**	**956,724.00**	**1,243,741.20**	**1,368,115.32**
Cost of goods sold	−2,254.00	−143,508.60	−287,017.20	−373,122.36	−410,434.60
Personnel expenses	−303,345.00	−306,985.14	−310,668.96	−314,396.99	−318,169.75
Other expenses	−460,572.00	−466,098.86	−471,692.05	−477,352.35	−483,080.58
Depreciation expense	−18,577.00	−417,175.00	−417,175.00	−417,175.00	−417,175.00
Operating profit/(Loss)	**−545,567.00**	**−855,405.60**	**−529,829.21**	**−338,305.50**	**−260,744.61**
Operating profit/(Loss) *(1-t)	−409,175.25	−641,554.20	−397,371.91	−253,729.13	−195,558.46
Depreciation Expense	18,577.00	417,175.00	417,175.00	417,175.00	417,175.00
Stocks variation	–	90,732.00	–	–	–
Customers variation	–	402,242.58	−39,317.42	−23,590.45	−10,222.53
Supplier Variation		−390,052.00	–	–	–
Cash Flow	**−390,598.25**	**−121,456.63**	**−19,514.33**	**139,855.42**	**211,394.01**

Table 3 Cash-Flows

Period (t)	CFt
0	−4,300,347.00
1	−390,598.25
2	−121,456.63
3	−19,514.33
4	139,855.42
5	211,394.01
Continuation	14,092,932.92

the liquidation of its debt, the project generates positive cash flows, equaling the free cash flow.

Once the cash flows have been estimated, the WACC must be calculated to update their value. This variable makes it possible to obtain the discount rate used to estimate the present value of the future cash flows of the company and reflects the cost of the resources employed, and those corresponding to debt and equity (Miller, 2009; Murray & Shen, 2016). The data and parameters used to obtain this value are the cost of capital, cost of debt, equity, debt and rates. The formula for obtaining the discount rate is as follows:

$$WACC = Rf + [E(Rm) - Rf] * \beta c \qquad (3)$$

The WACC formula is composed of the following variables: Rf es is the risk-free return, Rm is the market premium and βc represents the leveraged beta of the company. To determine the value of Rf, the 10-year Spanish bond indicator is taken at the date on which the calculation was made (September 2021) (Cinco Días, 2021). For its part, the Rm variable receives the value of the September 2021 market premium, and for the calculation of βc the Damodaran (2020) work was used, in which the value of the betas for the different business sectors is characterized. To deleverage the beta of the company analyzed, the deleveraged sector beta is adopted as a reference and the weight of the financial structure of the company (equity and borrowed funds) is added. Once the formula has been applied, the WACC value is 1.5%.

After determining the WACC, the NPV and IRR are calculated to assess whether the project is viable or not. While the NPV estimation provides the present value of the project, the IRR shows the return on investment and represents the discount rate at which the NPV is equal to 0 (Ho & Lee, 1986). This indicator assumes that, if the IRR is higher than the discount rate, the project should be accepted or carried out (Krüger et al., 2015). If, on the other hand, the IRR is lower than the discount rate, it is not advisable to carry out the project. In addition, the IRR based on free cash flows measures the profitability of the project without considering how it has been financed and, in turn, estimates the profitability for the shareholder (Ball et al., 2016; Estrada, 2011).

Table 4 Calculation of NPV
and IRR

Valuation model	Results
NPV	8,394,851.96 €
IRR	20.65%

Table 5 Sensibility analysis

Sales growth	NPV (€)	IRR (%)
Base case	8,394,851.96	20.65
Case 1	19,214,998.96	33.6
Case 2	31,215,166.72	43.15

The NPV of the project has been calculated by applying Eq. (1). The result obtained is €8,394,851.96. Therefore, it can be concluded that it is profitable to invest in this project (Table 4). In addition, the calculation of the IRR through Eq. (2) gives a return of 20.65%. This return is much higher than the discount rate (1.5%). Therefore, evaluating the result of the two models, it is concluded that the investment in a high value-added fish processing plant is profitable.

In addition to the valuation of the project, a sensitivity analysis was performed. For this purpose, three different scenarios were considered in order to forecast the economic evolution of the company depending on the behavior of the different economic factors: (1) base scenario; (2) 15% increase in growth in years 4 and 5; (3) 30% increase in growth in years 4 and 5.

Table 5 provides the results obtained after the application of the sensitivity analysis. The effect of sales growth in the fourth and fifth years is very significant in the valuation. In case 1 the increase in sales has led to a doubling of the NPV. As higher growth is assumed (case 2) the value of the investment is highly elevated.

5 Discussion and Conclusions

Spanish frozen fish companies, and more specifically the Galician ones, have evolved toward a new business model for the commercialization of their products. This model is based on the production and distribution of frozen products with a higher level of quality and sustainability. An example of this new paradigm in the Spanish frozen food sector is all those precooked dishes with seafood and fish.

This production model is based on efficiency and the possibility of offering a high value-added product to the market. These companies adopt the Lean Manufacturing system to optimize their resources and improve the quality of their service. In this way, the aim is to position, or consolidate, the firm's position in the market.

The company analyzed in this study is a leading company in the Spanish frozen food industry. The creation of a new processing plant, with its own brand name, is primarily focused on entering a new market niche: high-quality frozen products.

Premium frozen products have a high nutritional value and optimum organoleptic characteristics for the consumer. These products, within the food sector, have been called "the fifth range of Gourmet products", which, initially, is targeted at the Horeca sector but which, in the future, will be marketed in food distribution chains in order to expand its consumption to households.

The main problems that businesses face when starting this type of activity are, on the one hand, the need to make a high investment and, on the other hand, the shortage of highly qualified human resources necessary to face an activity with a high degree of innovation. This is due to the fact that traditional products are marketed in an industrial way under high-quality standards. In particular, the company in our case study requires a very high initial investment. This fact will generate losses for the company during the first years of activity, specifically during the first 5 years. However, after performing the analysis using the IRR methodology and considering a conservative sales figure, it is observed that positive cash flows are generated from the fourth year of activity onwards. Thus, it is concluded that, once the initial investment has been discounted and the cash flows generated have been discounted using the appropriate discount rate (WACC), the NPV of the project will be positive and will reach €8,394,851.96, with an IRR of 20.65%.

The analysis of the evolution of the results of these two models (NPV and IRR) demonstrates the viability of the project. Thus, it is concluded that fish processing in a sustainable and efficient way is not only environmentally responsible, but it is also an activity in which high profitability is obtained.

In addition to assessing this project from the point of view of its feasibility, we consider it appropriate to carry out a sensitivity analysis. For this purpose, three different scenarios have been designed. The first one is the base scenario, which is calculated using the variables included in the previous calculations and would yield an NPV of €8,394,851.96 and an IRR of 20.65%. In the second scenario, there is a growth in the fourth and fifth years of 15% in sales, which would lead to an NPV of €19,214,998.96 and an IRR of 33.6%. Finally, in the third scenario, sales growth for years 4 and 5 is 30%, leading to an NPV of €31,215,166.72 and an IRR of 43.15%. That is, in this third scenario the IRR would double that obtained in the base year.

The results shown above describe a scenario in which it is financially profitable for the company analyzed to continue with the new project. This project of diversification of the activity of the firm, in addition to being economically viable, is also viable in the social aspect. This statement is based on the fact that the production model implemented is more sustainable and efficient than those used by other companies in the sector. Furthermore, the high degree of R&D&I not only reduces costs or increases the added value of the products, but also helps to improve the corporate image of the enterprise, positioning it as a benchmark in the market.

Acknowledgements The authors would like to thank Juan Carlos López Rodríguez for his contributions to this article. This study was possible thanks to the financial support from Xunta de Galicia (ED431C2018/48 and ED431E2018/07) and from the Ministry of Economy and Competitiveness (RTI2018-099225-B-100). Also, Raquel Fernández-González thanks for financial support of the

Postdoctoral Program Xunta de Galicia under grant ED481B2018/095 and Raisa Pérez-Vas grate-fully acknowledges funding under grant ED481A-2018/341 from the Programa Predoctoral Xunta de Galicia.

References

Ardan. (2020). *Informe Ardan Galicia 2020*. Consorcio de zona Franca de Vigo.

Arrow, K. J., & Kruz, M. (2013). *Public investment, the rate of return, and optimal fiscal policy*. RFF Press.

Autoridad portuaria de Vigo. (2021). *Blue Growth Atlantic Vigo—Proyectos*. http://bluegrowthvigo. eu/proyectos. Accessed 21 September 2021.

Ball, R., Gerakos, J., Linnainmaa, J. T., & Nikolaev, V. (2016). Accruals, cash flows, and operating profitability in the cross section of stock returns. *Journal of Financial Economics, 121*(1), 28–45. https://doi.org/10.1016/j.jfineco.2016.03.002

Bello Bugallo, P. M., Cristóbal Andrade, L., Magán Iglesias, A., & Torres López, R. (2013). Integrated environmental permit through Best Available Techniques: Evaluation of the fish and seafood canning industry. *Journal of Cleaner Production, 47*, 253–264. https://doi.org/10.1016/j.jclepro.2012.12.022

Block, S. (2011). Does the weighted average cost of capital describe the real-world approach to the discount rate? *The Engineering Economist, 56*(2), 170–180. https://doi.org/10.1080/0013791X. 2011.573618

Bode, A., Alvarez-Ossorio, M. T., Cabanas, J. M., Miranda, A., & Varela, M. (2009). Recent trends in plankton and upwelling intensity off Galicia (NW Spain). *Progress in Oceanography, 83*(1), 342–350. https://doi.org/10.1016/j.pocean.2009.07.025

Borgonovo, E., Gatti, S., & Peccati, L. (2010). What drives value creation in investment projects? An application of sensitivity analysis to project finance transactions. *European Journal of Operational Research, 205*(1), 227–236. https://doi.org/10.1016/j.ejor.2009.12.006

Bureau Van Dijk Electronic Publishing group. (2021). *Orbis | Company information across the globe | BvD*. https://orbiseurope.bvdinfo.com/version-2021923/orbis4europe/Companies/Login?return Url=%2Fversion-2021923%2Forbis4europe%2FCompanies. Accessed 28 September 2021.

Business Wire. (2021). *Frozen food market by product type and user: global opportunity analysis and industry forecast, 2020–2027*. https://www.businesswire.com/news/home/20200%208250 05521/en/Global-Frozen-Food-Market-2020-to-2027---by-Product-Type-and-User---Resear chAndMarkets.com. Accessed 20 September 2021.

Carlson, A., & Frazão, E. (2014). Food costs, diet quality and energy balance in the United States. *Physiology & Behavior, 134*, 20–31. https://doi.org/10.1016/j.physbeh.2014.03.001

Chakrabarti, A., & Zeaiter, H. (2014). The determinants of sovereign default: A sensitivity analysis. *International Review of Economics & Finance, 33*, 300–318. https://doi.org/10.1016/j.iref.2014. 06.003

Chen, J., Jayachandran, M., Bai, W., & Xu, B. (2022). A critical review on the health benefits of fish consumption and its bioactive constituents. *Food Chemistry, 369*, 130874. https://doi.org/10. 1016/j.foodchem.2021.130874

Cinco Días. (2021). *Cotización y precio de Bono español*. Cinco Días. http://cincodias.elpais.com/ mercados/prima-de-riesgo/bono_espanol/1/. Accessed 28 September 2021.

D'.Amico, P., Nucera, D., Guardone, L., Mariotti, M., Nuvoloni, R., & Armani, A. (2018). Seafood products notifications in the EU Rapid Alert System for Food and Feed (RASFF) database: Data analysis during the period 2011–2015. *Food Control, 93*, 241–250. https://doi.org/10.1016/j.foo dcont.2018.06.018

Damodaran, A. (2020). *Equity risk premiums: Determinants, estimation and implications—The 2020 Edition*. Social Science Research Network.

Dusseault, B., & Pasquier, P. (2021). Usage of the net present value-at-risk to design ground-coupled heat pump systems under uncertain scenarios. *Renewable Energy, 173*, 953–971. https://doi.org/10.1016/j.renene.2021.03.065

Estrada, J. (2011). NPV and IRR. In J. Estrada (Ed.), *The essential financial toolkit: Everything you always wanted to know about finance but were afraid to ask* (pp. 116–135). Palgrave Macmillan UK.

FAO. (2021). *Spain is the European leader in the production and consumption of frozen seafood products | GLOBEFISH |*. Food and Agriculture Organization of the United Nations. http://www.fao.org/in-action/globefish/fishery-information/resource-detail/es/c/338579/. Accessed 20 September 2021.

Fernández-González, R., Pérez-Pérez, M.,& Garza-Gil, M. D. (2020). An analysis of production factors for Galician-farmed turbot: From boom to stagnation. *Aquaculture Economics & Management*, 1–19. https://doi.org/10.1080/13657305.2020.1840659

Fernández-González, R., Pérez-Pérez, M. I., & Garza-Gil, M. D. (2021). Main issues and key factors for development of turbot aquaculture in Spanish regions: A social-ecological perspective. *Aquaculture, 544*, 737140. https://doi.org/10.1016/j.aquaculture.2021.737140

Fuentes, A., Masot, R., Fernández-Segovia, I., Ruiz-Rico, M., Alcañiz, M., & Barat, J. M. (2013). Differentiation between fresh and frozen-thawed sea bream (Sparus aurata) using impedance spectroscopy techniques. *Innovative Food Science and Emerging Technologies, 19*, 210–217. https://doi.org/10.1016/j.ifset.2013.05.001

Garza-Gil, M. D., Surís-Regueiro, J. C., & Varela-Lafuente, M. M. (2017). Using input–output methods to assess the effects of fishing and aquaculture on a regional economy: The case of Galicia, Spain. *Marine Policy, 85*, 48–53. https://doi.org/10.1016/j.marpol.2017.08.003

Garza-Gil, M. D., Varela-Lafuente, M., & Caballero-Miguez, G. (2009). Price and production trends in the marine fish aquaculture in Spain. *Aquaculture Research, 40*(3), 274–281. https://doi.org/10.1111/j.1365-2109.2008.02106.x

Grand View research. (2021). *Frozen Food Market Size, Share & Trends Report, 2020–2027*. https://www.grandviewresearch.com/industry-analysis/frozen-food-market. Accessed 20 September 2021.

Ho, T. S. Y., & Lee, S.-B. (1986). Term structure movements and pricing interest rate contingent claims. *Journal of Finance, 41*(5), 1011–1029. https://doi.org/10.1111/j.1540-6261.1986.tb02528.x

Industrias Pesqueras. (2021). *El tráfico de pesca congelada del puerto de Vigo creció un 1,16 % en 2020*. https://industriaspesqueras.com/noticia-64450-seccion-Puertos. Accessed 21 September 2021.

Krüger, P., Landier, A., & Thesmar, D. (2015). The WACC Fallacy: The real effects of using a unique discount rate. *Journal of Finance, 70*(3), 1253–1285. https://doi.org/10.1111/jofi.12250

La Voz de Galicia. (2021). *Iberconsa reforzará su departamento de I+D+i*. Voz Galicia. https://www.lavozdegalicia.es/noticia/somosmar/2021/03/10/iberconsa-reforzara-departamento-idi/00031615405410526527394.htm. Accessed 21 September 2021.

León-Bravo, V., Moretto, A., Cagliano, R., & Caniato, F. (2019). Innovation for sustainable development in the food industry: Retro and forward-looking innovation approaches to improve quality and healthiness. *Corporate Social Responsibility and Environmental Management, 26*(5), 1049–1062. https://doi.org/10.1002/csr.1785

Li, N., Wu, X., Zhuang, W., Xia, L., Chen, Y., Wu, C., Rao, Z., Du, L., Zhao, R., Yi, M., Wan, Q., & Zhou, Y. (2020). Fish consumption and multiple health outcomes: Umbrella review. *Trends in Food Science & Technology, 99*, 273–283. https://doi.org/10.1016/j.tifs.2020.02.033

Miller, R. A. (2009). The weighted average cost of capital is not quite right. *The Quarterly Review of Economics and Finance, 49*(1), 128–138. https://doi.org/10.1016/j.qref.2006.11.001

Murray, M. Z., & Shen, T. (2016). Investment and the weighted average cost of capital. *Journal of Financial Economics, 119*(2), 300–315. https://doi.org/10.1016/j.jfineco.2015.09.001

Nussinovitch, A., & Peleg, M. (2000). Analysis of the fluctuating patterns of microbial counts in frozen industrial food products. *Food Research International, 33*(1), 53–62. https://doi.org/10.1016/S0963-9969(00)00023-5

Peymankar, M., Davari, M., & Ranjbar, M. (2021). Maximizing the expected net present value in a project with uncertain cash flows. *European Journal of Operational Research, 294*(2), 442–452. https://doi.org/10.1016/j.ejor.2021.01.039

Semeano, A. T. S., Maffei, D. F., Palma, S., Li, R. W. C., Franco, B. D. G. M., Roque, A. C. A., & Gruber, J. (2018). Tilapia fish microbial spoilage monitored by a single optical gas sensor. *Food Control, 89*, 72–76. https://doi.org/10.1016/j.foodcont.2018.01.025

Sen, S., Antara, N., & Sen, S. (2021). Factors influencing consumers' to Take Ready-made Frozen Food. *Current Psychology, 40*(6), 2634–2643. https://doi.org/10.1007/s12144-019-00201-4

Trippi, R. R. (1989). A discount rate adjustment for calculation of expected net present values and internal rates of return of investments whose lives are uncertain. *Journal of Economics and Business, 41*(2), 143–151. https://doi.org/10.1016/0148-6195(89)90013-1

United Nations. (2021). *Sustainable Development Goals*. U. N. Sustain. Dev. https://www.un.org/sustainabledevelopment/. Accessed 20 September 2021.

Vartiainen, E., Masson, G., Breyer, C., Moser, D., & Román Medina, E. (2020). Impact of weighted average cost of capital, capital expenditure, and other parameters on future utility-scale PV levelised cost of electricity. *Progress in Photovoltaics: Research and Applications, 28*(6), 439–453. https://doi.org/10.1002/pip.3189

Yakovleva, N., & Flynn, A. (2004). Innovation and sustainability in the food system: A case of chicken production and consumption in the UK. *Journal of Environmental Policy and Planning, 6*(3–4), 227–250. https://doi.org/10.1080/1523908042000344096

Žižlavský, O. (2014). Net present value approach: Method for economic assessment of innovation projects. *Procedia—Social and Behavioral Sciences, 156*, 506–512. https://doi.org/10.1016/j.sbspro.2014.11.230

Index

A
Abilities, 47, 50, 54
Access, 50, 54, 55
Across borders, 44
Adaptation, 42, 43
Affective commitment, 10, 15
Against, 109, 114–116
Agenda 2030, 111
Animal food waste, 122, 126–130, 138,
    144, 145
Anti-discriminatory selection process, 47
Artificial intelligence, 62
Attitudes, 48, 52, 55
Attraction-selection-friction model, 52
Austria, 68, 73, 76

B
Belgium, 73, 74, 76
Biases, 43, 49, 50, 52
Borders, 22
Brazilian, 108, 110, 114, 116
Brazilian environmental public policies,
    107
Breach, 6–8, 11–14
Bridging, 32
Broader terms, 11
Bulgaria, 74, 85, 95
Business, 22–25, 29–31, 157
Business human rights, 25

C
Candidates, 45, 47, 49, 50, 52, 55
Challenges, 22, 43
Chicago Model, 23

Citizens, 63, 66–68, 75–81, 83, 84, 86, 87,
    89, 91, 92, 94, 95, 98, 99, 107
Co-creation, 107
Collaborators, 42, 46
Commitments, 153
Common end, 21
Communication, 2, 3, 26, 30–33
Communication activities, 30
Communication theory, 30
Competences, 25, 32
Competition, 64
Competitive environment, 108
Compliance-based strategy, 29
Consumers, 120, 121, 125
Contents, 6, 11–13
Corporate social responsibility (CSR), 2, 3,
    14
Corporate Social Responsibility Human
    Resources Management
    (CSRHRM), 30
Corporate Sustainability (CS), 22
COVID-19 pandemic, 109, 114–116
Croatia, 75
CSR concerns, 24
CSR initiatives, 28, 30
CSR-orientation, 2, 29
CSR-oriented communication, 2, 3, 8, 10
CSR-related statements, 2
CSR-talk, 8
CSR-values, 2
CSR-washing, 2, 3
Cultural adjustment, 49
Cyprus, 75, 76, 95
Czechia, 76, 77

D
Data, 84, 86–93, 98, 99
Data as assets, 66
Data collection, 63, 64
Data management, 62–64, 98
Decent work, 22
Decision-making processes, 29
Democratic systems, 108
Demographic pressures, 43
Denmark, 77, 79, 80, 83, 95, 99
Description, 48
Designing, 31
Development, 6, 8, 10, 12, 42, 44, 48, 49
Differentiation, 46
Digital behavioral typologies, 63, 68, 69
Digitalization, 62, 64
Digital platforms, 63, 64, 98
Digital safety, 64
Dimensions of sustainable development, 62
Discrimination, 42, 47, 54
Displacements, 43
Diverse talents, 48
Diverse workforce, 43, 46, 47, 54
Diversification, 151, 157
Diversity, 42, 43, 46, 47, 52
Dual labor market theory, 45

E
Ecological goals, 23
Economic, 62, 64, 66, 108, 112, 114, 115
Economic activities, 121–123, 126, 130,
     135, 136, 141, 143, 145
Economic data, 152
Economic development, 122, 145
Economic growth, 22
Ecosystems, 110, 112
Educational qualifications, 45, 46
Efficiency, 150, 156
Electronic waste, 62
Employee(s), 2, 3, 5–15
Employee-organization dyad, 11
Employee retention, 2, 15
Employee well-being, 15
Employers-employees, 34
Employment, 1–3, 5–9, 12, 15
End, 24
Engaging, 63, 64, 89, 95
Enron, 29
Entry-level jobs, 5, 6, 8
Environmental, 62
Environmental balance, 108
Environmental degradation, 108, 150

Environmental impacts, 62
Environmental justice, 107, 112–114, 116
Environmental pressures, 43
Environmental protection agency (EPA),
     122, 124
Environmental public policies, 114, 116
Environmental regression, 108
Environmental sustainability, 108, 109
Estonia, 78
Ethics, 25, 27, 29, 33, 35
Ethnicity, 42, 46, 52
Ethnic minority colleagues, 10
EU-27, 63, 66–68, 73–95, 98, 99
European countries, 75, 89, 122, 126–128,
     132, 142
European Waste Classification for Statistics
     (EWC-Stat), 122
Eurostat, 122–126, 128, 131, 134, 137, 138,
     140, 142, 144, 145
Exploitative working conditions, 25

F
Facebook, 63, 64, 66–68, 73–81, 83–95, 98,
     99
Facebook data, 66, 73–83, 85–95, 98
Features, 10, 11
Financial analysis, 151, 153
Financial data, 152
Financial goals, 23
Finland, 79, 80, 83, 95, 99
Food and Agriculture Organization of the
     United Nations (FAO), 120–122,
     125, 126, 144
Food availability, 121
Food loss, 120–122, 124–126, 130, 135,
     141
Food loss and waste database (FLW), 125,
     144
Food loss indices (FLI), 126, 144
Food need, 121
Food quality loss, 121
Food quality waste, 121
Food resource, 150
Food supply chain, 120, 121, 124
Food waste, 120–122, 124–130, 132, 138,
     144, 145
Forced, 43–45
France, 80
Freelancers, 9
Frozen, 149–152, 156, 157
Fulfillment, 7, 11–14
Functions, 45, 48, 49

G
Gender, 42, 46, 52
Germany, 80, 81, 84, 86
Global challenge, 21
Global CSR strategy, 24
Global Food Loss Index (GFLI), 126
Globalization, 43, 44
Global strategy, 111, 116
Governmental visions, 110
Greece, 81, 82
Green attitudes, 24, 26
Green behaviours, 26
Green competences, 26
Green HRM, 24–27, 33, 35
Green results, 26
Greenwashing, 2

H
Harvard Model, 23
High quality, 3
High-quality relationship, 15
Hiring immigrants, 43
Households, 121, 122, 124, 126–129, 132,
      133, 138, 139, 144, 145
HRM functions, 23, 32, 35
HRM practices, 25, 27, 29–32
HRM strategy, 29
Human Resource Manangement (HRM),
      22–27, 29, 30
Human resource practices, 6, 12, 14, 15, 42
Human resources, 3, 5, 6, 8, 14, 15
Human resources strategies, 47
Human rights, 108–110, 112
Hungary, 82

I
IC strategy design, 31
Immigrants, 42, 45–47, 50, 52–55
Immigrants' access, 50, 54, 55
Immigrant workforce, 45
Immigration, 43–45
Improvement, 14
Inadmissible regression, 109
Inclusion, 110, 116
Income generation, 108
individual worker performance, 33, 34
Influence, 43, 49, 50, 52, 55, 107, 116
Information and communication
      technologies (ICT), 64
Innovation, 150, 152, 157
Insecurity, 25
Integration, 42, 44, 108, 110, 114

Integration of migrants, 44
Intelligent data management, 62–64
Intelligent technologies, 62
Interactionist, 22, 25, 26, 30, 31, 33, 35
Interactionist approach, 21
Interactionist component, 25, 30, 35
Interest rate of return, 153
Internal Communication (IC), 22, 26,
      30–35
Internal rate of return (IRR), 153, 155–157
In-work poverty, 25
Ireland, 83
Italy, 84

J
Job analysis process, 47
Job satisfaction, 6, 10

K
Knowledge, 25, 26, 32, 33, 35, 47, 48, 50,
      54

L
Labor market, 8, 42, 43, 45–47, 50, 52, 54,
      55
Latvia, 85, 86
Level, 62, 63, 66–68, 78, 95, 98, 99
LGBTQ community, 9
List of waste (LoW), 122
Lithuania, 86, 87
Localizing, 111, 116
long-term attention, 25
Luxembourg, 86–88

M
Malta, 87, 88, 90
Management, 3, 5, 6, 8, 14, 15
Market niche, 156
Market segmentation, 66
Means, 24
Measurement, 121, 122, 124
Migration, 42–45
Migratory processes, 42–44, 46
Mixed food waste, 122, 126–131, 138, 144,
      145

N
NACE Rev. 2, 123, 124, 126
Netherlands, 88
Net present value (NPV), 153, 155–157

Network online platforms, 64, 65
Non-human life, 114
Non-regression, 110

O
"Old wine in new bottles", 22, 25, 33
Organizational context, 46, 54, 55
Organizational culture, 30
Organization competencies, 23
Organizations sustainable development, v
Output, 23, 25–29
Over-consumption, 121
Over-fulfillment, 7, 13, 14
Overlaps, 30, 32
Over-nutrition, 121

P
Participation, 107, 108, 110, 111, 113, 114, 116
Participation process, 107
Performance, 22, 23, 27, 29, 30, 32, 34
Personal data, 63, 64, 66, 98
Perspective, 22, 25, 26, 29–35, 50, 64, 68, 75, 99, 114, 150
Poland, 89, 91
Population structure, 63, 68
Portugal, 90
Post-pandemic era, 116
Pragmatic perspective, 25, 31
Predict, 63, 95, 98, 99
Preference, 63, 68, 75–80, 83, 94
Primary labor market, 45
Processes, 42, 43, 47, 49, 50, 53–55
Production chain, 150
Production model, 156, 157
Productivity, 5
Pro-environmental behaviours, 27
Professional experience, 45, 47
Profile, 48, 49, 52, 55
Profiling, 64, 66
Profitability, 153, 155, 157
Pro-social orientation, 11
Protection of minorities, 110
Psychological adaptation, 43
Psychological contract, 1, 2, 5–15
Public awareness, 113
Public policies, 108, 113, 114, 116
Public power, 111, 113

Q
Qualified human resources, 157

Quality of life, 62, 98

R
Race, 42, 46, 52
Recruitment, 42, 43, 47–50, 52–55
Recruitment process, 42, 43, 47, 48, 50, 55
Regional food loss index (RFLI), 126
Regression, 110, 114–116
Relational features, 11
Relationships, 2, 3, 5–8, 14, 15, 43, 45, 107, 108, 111, 112, 116
Research unicorn, 22, 35
Resource-Based View (RBV Theory), 23
Romania, 90, 91, 95

S
SDG 8, 42, 43
SDG 10, 42, 43
Secondary labor market, 45, 46, 54
Segmented labor market theory, 45, 46
Selection, 43, 47–50, 52–55
Selection criteria, 49
Selection process, 47–50, 52
Self-employed, 6, 8, 9
Self-interest, 11
Shareholder, 153, 155
Short-term relationship, 10
SHRM practices, 33
Similarity, 49, 52–55
Similarity-attraction theory, 43, 50, 52, 54
Similar wastes, 122, 126, 138, 139, 141, 143–145
Skills, 26, 32, 42, 43, 45–50, 54
Slovakia, 91, 92
Slovenia, 92, 93
Social, 62, 66, 82, 85
Social adaptation, 43
Social exchange, 7, 13, 14
Social goals, 23
Social identity theory, 43, 50, 54
Socially responsibility, 25
Social networks, 63–65, 67, 78, 80, 87, 89–92, 98, 99
Social participation, 107, 108, 110, 112, 116
Social transformation, 44
Socio-economic imperative, 62
Socioeconomic vulnerability, 112
Spain, 93
Spanish frozen food industry, 156
Specification, 48, 49
Stakeholders, 5, 10, 14, 107, 109

Strategic framework, 30
Strategy, 22–25, 29, 31, 32
Sustainability, 21–27, 30–35, 62, 63, 67,
    68, 74, 76, 79–88, 91, 93, 94, 98, 99,
    108, 109, 114–116, 150, 156
Sustainability-related content, 63, 66, 68,
    80, 89–92, 98, 99
Sustainable, 1–3, 5–8, 10, 12, 14
Sustainable against regression, 107
Sustainable concerns, 29
Sustainable development, 62–64, 66–68,
    74–76, 78, 90, 95, 98, 99, 121, 122
Sustainable Development Goals (SDGs),
    22, 42, 62, 63, 67, 95, 108, 109, 111,
    116, 119, 121, 125, 150
Sustainable employment relationship, 2, 3
Sustainable environment, 3
Sustainable Human Resource
    Manangement (SHRM), 22–26,
    28–31, 33–35
Sustainable hype, 2
Sustainable management, 1–3, 5
Sustainable path, 29
Sustainable process, 150
Sustainablity, 25
Sustainablity management, 25
Sustained employment, 15
Sweden, 94, 95, 99
Symmetric communication model, 31

T
Talk, 2, 10
Targeted advertising, 66

Tilburg Psychological Contract
    Questionnaire (TPCQ), 12
Transactional feature, 10
Trigger discrimination, 54
Triple bottom line, 24

U
Under-fulfillment, 13, 14
Unemployment, 25, 42, 44
Unicorn, 25, 33

V
Vacancy, 48, 49
Value chain, 23
Value creation, 30
Vegetal wastes, 122, 124, 126, 145
Voluntary, 43
Vulnerabilities, 112, 116

W
Walk, 10
Waste measurement, 122
Waste statistics regulation (WStatR), 122
Well-being, 108, 114
Worker attitudes, 33, 34
Workers, 26, 29, 31–34
Worker's life cycle, 32
Workforce, 2, 3
Working environment, 47
Worldcom, 29

Ingram Content Group UK Ltd.
Milton Keynes UK
UKHW020146090523
421397UK00002B/4